Engineering Information
Management Systems

Automation In Manufacturing
Titles from Van Nostrand Reinhold

Engineering Information Management Systems

Beyond CAD/CAM to
Concurrent Engineering Support

John Stark

AUTOMATION IN MANUFACTURING SERIES

VAN NOSTRAND REINHOLD
New York

Library of Congress Catalog Card Number 91-36567
ISBN 0-442-01075-3

Printed in the United States of America

Van Nostrand Reinhold
115 Fifth Avenue
New York, New York 10003

Chapman and Hall
2-6 Boundary Row
London, SE 1 8HN, England

Thomas Nelson Australia
102 Dodds Street
South Melbourne 3205
Victoria, Australia

Nelson Canada
1120 Birchmount Road
Scarborough, Ontario M1K 5G4, Canada

16 15 14 13 12 11 10 9 8 7 6 5 4 3 2~1

Library of Congress Cataloging-in-Publication Data

Stark, John, 1948–
    Engineering information management systems— : beyond CAD/CAM, to
concurrent engineering support. / John Stark.
      p.   cm. — (Automation in manufacturing)
    Includes index.
    ISBN 0-442-01075-3
    1. Production engineering—Data processing. 2. Production
management—Data processing. 3. Data base management. I. Title.
II. Series.
  TS176.S752 1992
   670.'285—dc20
                                        91-36567
                                        CIP

Contents

Preface

Engineering information management (EIM) systems help:

- Reduce the time to introduce new products,
- Reduce the cost of developing new products,
- Reduce the cost of new products, and
- Improve the quality of products and services.

Engineering information management systems have multiple roles. They can be used to manage CAD data. They help improve the flow, quality, and use of engineering information throughout the company. They provide improved management of the engineering process through better control of data, engineering activities, engineering changes, and product configurations. They provide support for the activities of product teams and for advanced techniques such as concurrent engineering. Their multiple role is clearly apparent from the other acronym, EDM/EWM (engineering data management/engineering workflow management), that is also used to describe them. The acronym, EDM/EWM will be used throughout this book to reinforce the multiple objectives of EIM.

For companies with large engineering operations, and for those looking to gain a competitive advantage through their engineering activities, EDM/EWM will be a key technology in the 1990s. Managers of engineering companies need to be aware of and involved with EDM/EWM. As competitive pressures increase, lead times must decrease and product quality must increase. EDM/EWM systems can help achieve these goals. Since engineering information is used throughout the company, cross-functional implementation of EDM/EWM is necessary. Although successful implementation will require much time and money, the results will justify the investment. Due to its cross-functional, long-term, and high budget characteristics, EDM/EWM requires significant top management involvement.

Chapter 1 sets the scene for the book. It outlines the role and objectives of EDM/EWM systems. It describes the benefits that can be obtained from their use. From the experience of pioneers in the use of these systems, it appears that they can help:

- Reduce engineering costs by at least 10 percent,
- Reduce the product development cycle by at least 20 percent,

- Reduce engineering change handling time by at least 30 percent, and
- Reduce the number of engineering changes by at least 40 percent.

It is not easy to introduce a cross-functional system such as EDM/EWM. Many problems can occur. The causes of these problems are identified, and suggestions to overcome them are proposed. The need to sell the concept of EDM/EWM to many people, at many levels of the company, before starting implementation, is stressed.

Chapter 2 looks at some of the main EDM/EWM issues from the point of view of top management. The objectives of EDM/EWM are described from the management viewpoint. After defining "engineering data" and "engineering workflow," some of the reasons why EDM/EWM is becoming so important are addressed and it is shown why top management must become involved with it. EDM/EWM systems are a necessary foundation of a successfully integrated computerized engineering environment. To underscore the need for EDM/EWM, some of the potential benefits of its use are described.

Chapter 3 addresses the importance of recognizing the need to manage engineering workflow. It is only too easy to try to manage engineering data without taking account of the workflow. Workflow improvement techniques, such as just-in-time and concurrent engineering, that can lead to major benefits are described. The link between engineering data management and engineering workflow management is discussed. Unless the two issues are addressed together, expected benefits are unlikely to occur.

Chapter 4 describes the components of an EDM/EWM system that are needed to meet the requirements outlined in the first two chapters. These include an Information Warehouse, an Information Warehouse Manager, a Product and Workflow Structure Definition Module, a Workflow Control Module, and a Configuration Management Module. Other components are an Interface Module, a system administration module, and the basic infrastructure of a networked multi-vendor computer environment.

Chapter 5 describes in detail the need for EDM/EWM systems from the point of view of middle management and other potential users of these systems, such as design engineers, drafters, part programmers, process planners, shop floor workers, and maintenance engineers. EDM/EWM systems respond to a variety of current problems, ranging from the diversity and volume of engineering data to the difficulty of maintaining multiple versions, definitions, and representations of data. Other problems addressed include the difficulty to maintain product configurations and traceability, and the fact that data is often file based, multiply defined, and difficult to access. The detailed technical viewpoint adopted in Chapter 5 complements the top management view of Chapter 2.

Chapters 6, 7, and 8 look at some of the details of EDM/EWM from the viewpoint of data management. Some readers may prefer to read them after reading Chapters 9 and 10, which address the implementation of EDM/EWM.

The need for data base management systems to address the large volumes of engineering information is addressed in Chapter 6. It is shown why commercial data base systems are generally inappropriate for the engineering environment. Possible solutions are discussed. The hierarchical, network, and relational data models are briefly described, as are the more recent developments in extended relational data bases and object-oriented data bases.

Chapter 7 addresses the need to model the engineering process, as well as the flow and use of engineering information. It shows how this can be done, and outlines data flow, control flow, data structure, and control structure modelling techniques.

Data exchange issues and standards for data exchange are introduced in Chapter 8. Topics such as EDI, OSI, and IGES are discussed. The importance of the US DoD's CALS program, a major factor behind the recent rapid development of EDM/EWM, is stressed. The role of PDES/STEP is described.

Chapters 9 and 10 address the process of implementing an EDM/EWM solution. They describe the many activities that must be carried out before EDM/EWM can be used successfully and the expected benefits achieved.

Appendix 1 describes the various types of system currently proposed to meet EDM/EWM needs. These include engineering data management systems, engineering drawing management systems, and engineering document management systems. Other systems, such as extended CADCAM systems, configuration management systems, and technical publishing systems, also have important roles to play.

Appendix 2 provides a brief overview of some 50 systems that can contribute to an EDM/EWM solution. These include systems from suppliers such as Control Data, DEC, EDS, IBM, Prime Computer, Tandem Computers, and Unisys. There are systems from smaller companies such as ACCESS, Eyring, FORMTEK, and Sherpa that address specific areas of engineering data, document, and drawing management. CADCAM vendors such as Adra, Auto-trol, Cimlinc, Intergraph, and Matra Datavision also provide systems.

Appendix 3 and Appendix 4 provide a brief guide to EDM/EWM product names and to some of the common acronyms of the EDM/EWM world.

1

Selling the Engineering Information Management Message

1.1 GRABBING MANAGEMENT ATTENTION

There are computer systems coming to the market that help improve the flow, quality, and use of engineering information throughout a company. These systems provide improved management of the engineering process through better control of engineering data, of engineering activities, of engineering changes, and of product configurations. They provide support for the activities of product teams and for advanced organizational techniques such as concurrent engineering.

From the experience of pioneers in the use of these systems, it appears that they can help:

- Reduce engineering costs by at least 10 percent,
- Reduce the product development cycle by at least 20 percent,
- Reduce engineering change handling time by at least 30 percent, and
- Reduce the number of engineering changes by at least 40 percent.

Systems like these have a strong effect on competitiveness, market share, and revenues, and help to:

- Reduce the time to introduce new products,
- Reduce the cost of developing new products,
- Reduce the cost of new products, and
- Improve the quality of products and services.

These systems can be referred to as engineering information management (EIM) systems, for this is what they are, even though this name does not really describe what they do. A more appropriate name would be EDM/EWM systems (engineering data management/engineering workflow management). This name highlights the fact that these systems are not just managing engineering data, but are also managing the flow of work.

EDM/EWM systems do several things. They manage engineering data—all the data related to a product and to the processes used to design, manufacture, and support the product. Much of this data will be created with computer-based systems such as CAD, CAM, and CAE. EDM/EWM systems also manage the flow of work through those activities that create or use engineering data. They support techniques, such as concurrent engineering, that aim to improve engineering workflow.

At a time when manufacturing success increasingly depends on the speed with which a company develops new products, EDM/EWM systems have an important role to play.

EDM/EWM systems are sometimes referred to as product data management systems or product information management systems, but these terms overemphasize product data management and neglect the objectives of managing process data and managing the flow of work through the processes that use engineering data.

EDM/EWM systems treat engineering information as an important resource that is used by many functions in a company. They allow companies to get control of engineering information and to manage activities in several departments. In the long term, EDM/EWM systems will allow companies to get control of all their engineering information and to manage the overall engineering process. These characteristics set them apart from systems, such as CAD, that aim to improve the productivity of individual tasks in one functional area. Viewed as data processing systems, EDM/EWM systems go beyond individual application programs such as CAD and NC. Viewed as organizational tools, they go beyond individual approaches such as DFA (design for assembly) and project management systems.

EDM/EWM systems provide a backbone for the controlled flow of engineering information throughout the product life cycle. Other systems using engineering data, such as CAD, NC, process planning, MRP, and field service, will be integrated to this backbone.

EDM/EWM addresses issues such as control, quality, reuse, security, and availability of engineering data. Much as CAD, CAD/CAM, CAE, and MCAE were important in the 1970s and 1980s, EDM/EWM will be key to successful engineering in the 1990s. EDM/EWM offers important new functions for the engineering environment. It will help solve many of the problems that beset today's engineering environment and, for those who master it, will offer new strategic opportunities.

The introduction of an EDM/EWM system will reveal many problems associated with the use of engineering information. However, the primary aim of an

EDM/EWM system is to manage engineering data and the associated workflow, not to resolve all the problems that occur in the engineering department, as well as any others that may contribute to unnecessarily long product development cycles. Other actions will have to be taken to reduce these problems, such as the introduction of concurrent engineering. The system can support these actions, but cannot automatically solve the problems.

Implementation of EDM/EWM will be a challenge. In the manufacturing world, rewards come to companies who have a little vision, pick the right strategy, organize accordingly, and then put in a lot of work. This book aims to help those companies that want to implement EDM/EWM successfully. Initially, these will be medium to large (e.g., Fortune 1000) companies. They will include engineering and construction companies; discrete manufacturers in automotive, aerospace, electronic, and mechanical engineering industries; and batch processing companies in the food and pharmaceutical sectors. These companies know that it is not easy even to successfully implement individual applications such as CAD. They are unlikely to underestimate the difficulty of implementing a cross-functional system such as EDM/EWM, which will take many years to implement and will have considerable technological and organizational impacts.

Small to medium companies can also benefit from engineering information management. Compared to the larger companies, they should find it easier to implement, since they will generally have less people involved and less data to manage.

1.2 THE OBVIOUS BENEFITS OF EDM/EWM

For those who are reading this because they already want to implement EDM/EWM, its benefits are obvious. In their companies, they realize that engineering data is out of control, engineering workflow is out of control, and corrective action is needed. For those who are not so sure, more explanation may be necessary.

The term "engineering data" includes all the data related to a product and to the processes that are used to design, produce, and support it. In the past, systems have not been available to manage all of this data, and so it has gone either unmanaged or only partly managed. This is not to say that companies have not used systems to manage any data. Typically, the EDP and the MIS departments or their successor, the IT or IS department, has implemented systems to manage "business data" such as sales data, financial results, corporate plans, manufacturing resource plans, and personnel information. However, the IT department will rarely have addressed the management of "technical data." Technical data will have been managed either on paper or within the computer systems, such as CAD, that have produced it.

Engineering data is difficult to manage because:

- There is a lot of it (with more being created each day),
- It is on many media (e.g., paper and magnetic disks),

- It is used by many people in different functions (often at different sites),
- It is used by many computer programs (often on different computers),
- It often has several (different) definitions,
- It exists in many different versions,
- It has multiple relationships and meanings, and
- It may need to be maintained for many years (e.g., 50 years).

The term "engineering workflow" refers to the flow of work through those activities that create or use engineering data. Engineering workflow is not limited to the flow of work through the engineering department. It also includes the flow through the other organizations that make use of engineering data. Some of these activities are carried out inside the company, others outside (e.g., by suppliers and customers).

There is a close link between engineering data and engineering workflow. Each step of the workflow makes use of data. Individual items or sets of data are often used in many steps of the workflow. The link between data and workflow implies that it is neither feasible nor efficient to address them separately, hence the need to address them together with an EDM/EWM system.

Typically, a company will become interested in EDM systems when it finds that it can no longer manage engineering data by traditional manual methods, (e.g., when it can't control all its CAD data). It will become interested in EWM when it senses that it has to reduce the time it takes to get a product to market. In traditional organizations, the workflow is sequential and the product development time is the sum of the times of individual workflow steps. Development time can be reduced by carrying out steps in parallel—that is, simultaneously.

EDM/EWM systems help:

- Reduce the time to introduce new products,
- Reduce the cost of developing new products,
- Reduce the cost of new products, and
- Improve the quality of products and services.

EDM/EWM systems help improve the flow, quality, and use of engineering information throughout the company. They provide improved management of the engineering process through better control of engineering data, engineering activities, engineering changes, and product configurations. They provide support for the activities of product teams and for advanced techniques such as concurrent engineering.

From the experience of pioneers in the field of EDM/EWM, it appears that EDM/EWM systems can help:

- Reduce engineering costs by at least 10 percent,
- Reduce the product development cycle by at least 20 percent,
- Reduce engineering change handling time by at least 30 percent, and
- Reduce the number of engineering changes by at least 40 percent.

These are very important benefits. For companies with large engineering organizations, and for those looking to gain competitive advantage through their engineering activities, EDM/EWM will obviously be a key technology in the 1990s.

1.3 SORRY, IT'S NOT OBVIOUS TO ME

Unfortunately for those wanting to implement EDM/EWM, there are many people in the company who will not understand why EDM/EWM is necessary. Among those who won't understand will be many whose support is essential to implementing any long-term, cross-functional, computer-based system that will have significant effects on company performance and organization.

Those who may have difficulty in understanding the need for EDM/EWM will probably include corporate management, corporate planners, top management, engineering management, and other engineering staff and IT professionals.

Their reactions to talk of EDM/EWM will include:

- "We're into quality these days, not technology." (corporate management)
- "We don't get involved in low-level issues." (corporate management)
- "It's CIM again. Everyone knows it doesn't work." (corporate planner)
- "It's just another cost. Don't you know we are in a cost limitation phase?" (financial controller)
- "If you can't show the payback period is less than 18 months, we're not interested." (corporate planner)
- "We design products, we're not computer freaks." (engineering management)
- "We have it. We bought CAD to do this ten years ago." (engineering management)
- "We have to get the product out the door. Go talk to engineering." (manufacturing management)
- "We've done it. We have a relational data base on our mainframe." (IT management)
- "This is the CAD manager's responsibility—not ours." (IT management)
- "We've heard it all before. It's only computer vendor hype." (many)
- "We've heard it all before. We don't believe in acronyms." (many)
- "It's early days for these systems. Come back in five years." (many)
- "We've had enough of computers. We're trying to simplify before automating." (many).

This is really the starting point of this book. EDM/EWM and EDM/EWM systems are necessary, but most companies will initially fail to understand this simple message, will not react to it in the right way, and will fail to obtain the potential benefits. This is not meant to be a negative message to the reader, just a realistic assessment of the likely situation. The objective of this book is to help the reader successfully implement an EDM/EWM system, in spite of the likely lack of interest.

Since EDM/EWM is initially beyond the grasp of many of the people in the company who should be its keenest supporters, the first problem for anyone interested in introducing EDM/EWM is not "how do I implement it?" but "how can I convince people they need it?" To convince people, the reader needs first to understand why they don't understand, and then to prepare the corresponding arguments.

1.4 UNDERSTANDING BEFORE IMPLEMENTING

The mountain facing anyone wanting to implement EDM/EWM (Figure 1.1) is not the "implementation mountain." It's the "understanding mountain."

Many of the people whose support is necessary and who control the resources needed for success just don't understand. Their level of understanding will be close to 0 percent, and it will take many months or even years to get it to a level where they will become supportive.

It's not possible to implement EDM/EWM without support throughout the company. It's not possible to go it alone. EDM/EWM is cross-functional. It is as much an organizational approach as a technological approach, and it needs positive involvement from many levels of people in many functions. A local, go-it-alone approach in one function will generate extra costs without providing the hoped-for

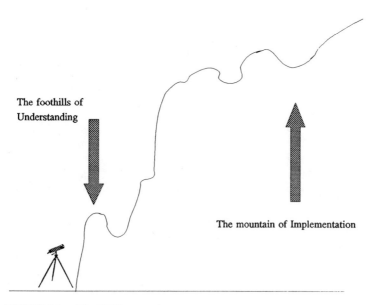

FIGURE 1.1.   The EDM mountain range

benefits. The only approach that will succeed in the long term is to fully understand the issues and convert the sceptics, and then implement across all functions. As a first step, it may be possible for one department to implement a prototype addressing one specific EDM/EWM problem, but this should only be seen as a local, short-term demonstration of potential capability.

1.5 MODELS OF INCOMPREHENSION

The exact reasons for the lack of understanding will differ from one company to company, but the basic causes are usually fairly similar. Some simple models will illustrate the problems.

The basic high-level value chain model (Figure 1.2) is slightly deceptive when used for purposes other then high-level positioning. It shows the product life cycle process (PLCP). The marketing function analyzes market requirements, and specifies products that the engineering function defines and that the manufacturing function produces. The sales function has responsibility for getting individual customers to buy individual products.

The model is deceptive for several reasons in that it implies there is only one "product" at the end of the process, it fails to show intermediary "products" along the chain, it ignores "products" that are used internally, and it appears to exclude product usage by the customer from the PLCP. Figure 1.3 expands the representation of the PLCP to show the many "products" produced during the process. Apart from the physical product, there are other products, such as customer perception and cost. More importantly for the subject of the book, there is also an information product. A great deal of information is generated to specify the product and the processes of the PLCP.

The importance of this information, generally excluded from value chain models, varies from industry to industry. In the development of new chemical entities in the pharmaceutical industry, the information is almost as important as the drug itself, since it has to be approved by the FDA before the drug can be marketed. In the aerospace industry, traceability is essential. In the automotive industry, if a car manufacturer doesn't maintain information on which air bag is fitted in which car, it may have to recall many more cars than necessary should an air bag fault occur.

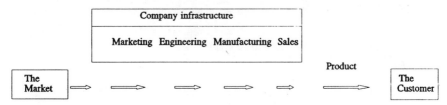

FIGURE 1.2   The basic value chain model

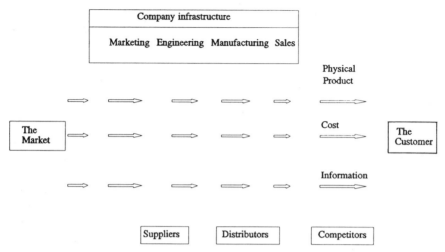

FIGURE 1.3   A detailed value chain model

If design information is available for reuse, it may be possible to reuse it and thus reduce design time. If it is not available, then tasks that have already been carried out once may have to be repeated, wasting valuable time. Information must be given its proper value and treated as a valuable corporate resource; otherwise, it soon becomes worthless. The reader will find many people in the company who underestimate the value of information.

The product life cycle process should be linked to corporate management and corporate planners (Figure 1.4). There should be good communications between the three groups. Corporate management should communicate plans, policies, and targets down the pyramid (Figure 1.5). The results of the process and proposals for improvement should be communicated up the pyramid.

Figures 1.4 and 1.5 are theoretically correct. However, they are rarely implemented in practice. Figures 1.6, 1.7, and 1.8 show the more usual situation. Figure 1.6 shows the typical organization chart with the functional empires defined. Often, as Figure 1.7 shows, the different functions are often physically separated as well as organizationally distinct.

Figure 1.8 illustrates both the functional and hierarchical barriers to communication.

1.6 CORPORATE MANAGEMENT—A DIFFERENT WAVELENGTH

In large corporations, corporate management clearly has a role to play. Focussed on the long term, it plays its textbook role of setting directions, objectives, strategies, organizational structures, plans, and budgets. Corporate managers also

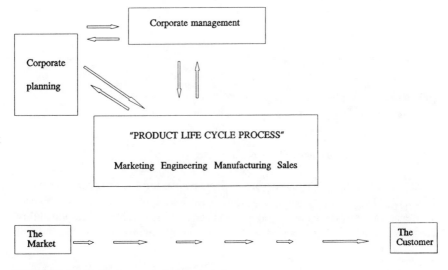

FIGURE 1.4   Corporate communication channels

play a role as leaders of the company, acting as figureheads for employees, shareholders, and customers.

Once corporate management has set the targets, the market plays its role, and corporate management has to wait to see the results of its chosen strategy. It can then decide whether to continue in the same direction or to change course. Provided that corporate management plays its textbook role, and either maintains its understanding of the product life cycle process or completely delegates the

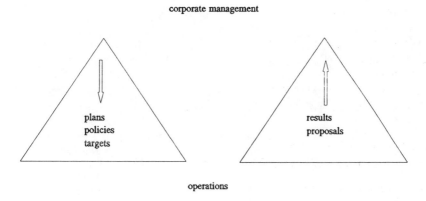

FIGURE 1.5   Communication up and down the corporate pyramid

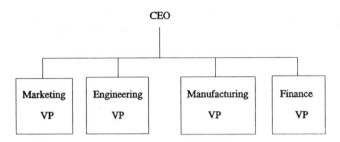

FIGURE 1.6   A typical functional organization chart

management of the PLCP, there is no intrinsic reason for it not to be supportive of EDM/EWM.

Recently, many large corporations have been shifting much of the responsibility for business unit strategies and objectives to the senior management of individual

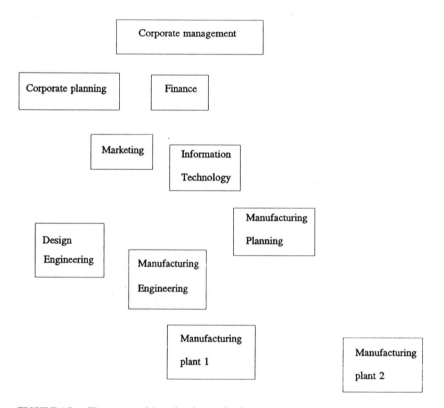

FIGURE 1.7   The separated functional organization

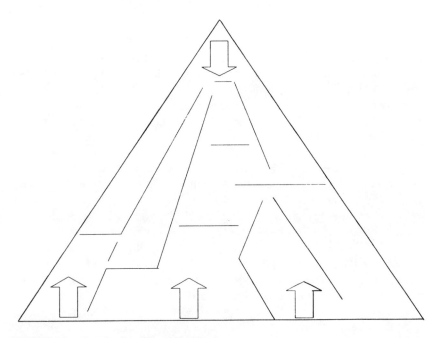

FIGURE 1.8   Functional and hierarchical barriers

business units, while focussing the corporate management role on overall capital, resource, and return management. One of the objectives of this switch is to give business unit responsibility to senior business unit managers who should understand their customers, products, and processes.

However, it is not infrequent for corporate management to fail to play its correct role or to fail to delegate sufficient responsibility to the business unit managers. In such cases, it often pays little attention to the product life cycle process, yet fails to clearly delegate PLCP responsibility. The resulting environment is unlikely to be favorable for the EDM/EWM debate. Similar problems can also occur in business units if senior managers play a traditional corporate management role, instead of the expected business unit management role.

Typical characteristics of such senior managers are that they deal with a very high-level abstraction of the business, have little understanding of the product life cycle process, have a poor relationship and little communication with the operating units, are very short term and dollar oriented, and love neither their customers nor their products. They do not take responsibility for their long-term role, yet do not clearly give the short-term responsibility to operating managers. They do not define the communication channels within the organization. They fail to provide operating managers with clear targets that do not contradict other managers targets.

"While our sector continues to experience below-average growth and we suffer from high interest rates and fluctuating exchange rates, we must be responsive to change as we enter a period of acquisitional growth. We will have to work much harder for profit growth in the next decade. To this end, we have implemented a 'Total Quality' program. We have continued to press the government to limit unfair competition from Japanese imports."

How does one get the type of corporate manager or senior business unit manager accustomed to pronouncing such motherhood messages to support implementation of EDM/EWM? Much of the message is devoted to world issues and market issues over which the company has little or no control. What about the issues over which the company does have control? Manufacturing, engineering, and information technology appear to be irrelevant to many senior managers. Often, it seems they have little practical experience of the product life cycle process, and instinctively shy away from anything that might show up their lack of knowledge of the details. They cannot afford to waste their time on low-level issues, preferring to focus on slogans such as "Total Quality," while leaving the product life cycle process to run itself.

There is nothing wrong with "Total Quality," but slogans of this type are meaningless to a company's workers unless they are translated into more concrete terms. "We will reduce our cost of quality by 10 percent." "We will halve product development times." Again, the messages sound wonderful, but have little meaning on the shop-floor or in the engineering department. "Cut inventories by 50 percent" and "reduce the number of engineering changes by 1500" have meaning for middle managers, but are still not detailed enough for those at the bottom of the pyramid, where they may even be contradictory. The potential implementer of EDM/EWM can run into trouble by trying to address the wrong targets.

If corporate and senior business unit managers' potential failures are dealt with at some length, it is because their support is so often essential to the success of cross-functional activities such as EDM/EWM. The reader has to understand the role that corporate managers are playing and if their support is needed, identify how to gain their backing.

1.7 CORPORATE PLANNING'S IVORY TOWER

Full implementation of EDM/EWM will take many years and affect many functions and people. It will cost a lot of money and should bring a lot of benefits. With these characteristics, it will have to be included in corporate budgets and investment plans. Anyone hoping to implement EDM/EWM had better get some support from the corporate planners. This will not be easy.

Many corporate planners have little hands-on experience of what happens throughout the product life cycle process, so its unlikely that they will instinctively support EDM/EWM. It may well be that their only job in the company will have been in the corporate planning department. They might even have joined it after getting their MBA. Probably, they won't have gone on any "Introduction to the Company" courses that are on offer. After all, they know all about the business world from their business school days, so they would have little to learn from such a course. As a result, they will have little idea of what the company is, who the customers are, what the products are, or the processes that a product goes through from its conception to its use by a customer.

Instead, they will have a spread-sheet view of the company—here are the costs, here are the sales, add some market share figures, some inflation figures, some market growth, some tax rates, and look! By year 3, we will have increased our profits. If we were now to merge our figures with those of Company Y, then we would increase our profits even more. Let's acquire Company Y. This kind of behavior, focussing on mergers, acquisitions, speculation, and asset trading, to the exclusion of operations, became very fashionable in the 1980s, generally to the detriment of operational performance.

In the 1990s, there appears to be a trend towards rotating seasoned business executives through the corporate planning department. Overall, this is a positive change. However, since these individuals have a wide range of backgrounds, ambitions, levels of enthusiasm, and willingness to get involved, as a group they do not show clearly defined behavioral characteristics.

It is often difficult to get traditional corporate planners to understand the actual difficulties involved in doing something or to understand the effects of something as difficult as defining the benefits of "better control of engineering information." These are head-in-the-clouds planners, not feet-on-the-ground doers. The seasoned business executives now coming into corporate planning departments are more likely to have their feet on the ground, but are often heavily influenced by past experience. However, if you want to implement EDM/EWM, you have to understand the corporate planners and find ways to get them to support the EDM/EWM project. This often means understanding their preferred plans and then aligning the benefits of EDM/EWM with these. It also means closely relating the expenses of EDM/EWM to improved corporate performance.

1.8 THE TOUGH LIFE OF THE TOP MANAGER

The top manager, running a functional department such as engineering or manufacturing, lives in the real world, and is expected to produce instant results. Talk to them about benefits of EDM/EWM appearing in two or three years time, and they will not show much interest. These managers are often so busy working at day-to-day problems that they cannot get involved in long-term issues. Their primary concern is to meet the short-term targets set for them by corporate management or by the CEO of the company.

At the corporate level, the right objectives may be set for the overall organization. Unfortunately, these are rarely translated into lower-level targets that fit together and reinforce each other. Of course, corporate managers and senior business unit managers want to reduce costs, reduce lead times, and improve quality, but they rarely take the trouble to translate these requirements into orchestrated targets for top functional managers. As a result, the top managers play as soloists. They don't have control of other functions, so they try to find solutions within their own departments. They will each react differently to a proposal for EDM/EWM.

A typical result of corporate management's intention to reduce costs, reduce lead times, and improve quality is that a VP of Quality will be recruited. This new function is set up almost as an independent department, although its objective can only be to improve quality in and between existing departments. The VP of Quality will be supportive of EDM/EWM, as it helps improve quality, but often, lacking a power base and real authority, may not be a key player.

The marketing department will react to corporate management's directive by identifying more finely segmented niche markets and the corresponding products. They can claim to be listening to the voice of the customer and, "like the Japanese," customizing the product line. At the same time they will grumble about lengthy product development cycles and the inability of the sales force to actually sell the products. Marketing is unlikely to be supportive of EDM/EWM, as they will probably see it as a tool to suppress their creative abilities.

The engineering department hears a mixture of messages. Corporate management is telling it to reduce costs, improve quality, and reduce lead times. At the same time, it has to meet the product development target dates in the corporate plan. It is faced with a marketing demand for many new products. As the individual requirements of the different requests are different or even contradictory, engineering has to walk a fine tight rope. It focusses on what it thinks is most important and most likely to be used by corporate management to measure its performance, and looks for ways to play down the other issues or to pass them on to others. Probably, there will be one or two key products that must be produced on time and to specification. Most of the effort will go on these. To show corporate management that serious steps are being taken, the engineering organization chart will be redrawn and some new operating procedures introduced. Marketing will be blamed for not providing full specifications, and for continually changing specifications, and manufacturing will be blamed for not being capable of producing the products that have been defined.

If all else fails, the engineering department can achieve the target of reducing lead times by releasing unfinished work to manufacturing. Of course, this will come back later in the form of rework, but since no one will have set a target for reduction of engineering changes, hopefully this will go unnoticed by corporate management. Involved daily in this environment, the Engineering VP may not be interested in the future benefits of EDM/EWM.

The Manufacturing VP will claim to have already set the objectives of reducing costs and improving manufacturing productivity, and cannot see how these targets can be achieved by better management of engineering data. After all, that should be the responsibility of the engineering department. This key user of engineering data is unlikely to be, at least in public, a strong supporter of EDM/EWM. In private, the Manufacturing VP may admit that anything that gives some control over engineering and engineering data cannot be wholly bad.

1.9 ENGINEERING'S OWN WORLD

In addition to those problems facing the Engineering VP, that are common to all top managers, the Engineering VP also faces some problems that are specific to the engineering department. Engineers are often highly educated, underutilized, not respected, and underpaid. For years, they do their job well and see little reward, whereas colleagues in other departments rise rapidly up the corporate hierarchy.

As functional organizations give them little chance to expand their skills horizontally by working in other departments, they eventually become isolated, highly focussed on their own engineering skills, and somewhat resentful of other departments. They see marketing as overindulged and overpaid, in view of the poor product specifications it produces. They look down on the poor old manufacturing function, which is incapable of handling even the simplest design. As for the bean counters in the finance department, who understand neither products nor processes, it is best to say nothing about them as, in spite of their apparent ignorance, they generally finish up at the top of the corporate tree. The IT function is seen as an appendage to the finance department, with absolutely no idea about the way to use computers in the engineering department.

With such a background, it may not be easy to convince the engineering department that EDM/EWM is either helpful or necessary. After all, they are, indisputably, highly skilled engineers, not computer programmers, and cannot see how a data management program can increase their skills. The potential implementer of EDM/EWM will need to find some good arguments.

1.10 THE IT VIEW

As EDM/EWM systems are computer-based, it might be expected that the IT function would support their introduction. However, this is rarely the case. It is more likely that the IT staff will never have worked with such systems, not understand what they are, and find it particularly difficult to address the workflow issues.

Initially, the IT group may try to claim EDM/EWM ownership by comparing them with data base systems used to support the F&A function, but it will soon be realized that there are too many differences. Another technological feature that may attract IT is the computer networking that will be needed to support communication of information.

In the past, the IT function has focussed on supporting the finance function, providing so-called "business systems." (It is interesting to note that the term "business systems" does not include computer systems concerned with engineering information. Apparently, users of engineering systems are not part of the business.) Since IT staff will rarely have any experience or understanding of the product life cycle process, it is unlikely that they will be strong supporters of EDM/EWM. On the contrary, they may even take a somewhat negative attitude, upset that people who are not IT specialists are proposing a new application for computers.

If the IT function can be convinced of the benefits of EDM/EWM, its close links to the Finance VP may provide a useful point of support for EDM/EWM. On the other hand, if they cannot be convinced, their lack of support may be interpreted by top management as meaning that EDM/EWM is a computer system of little value.

1.11 SELL BEFORE IMPLEMENTING

Before trying to implement EDM/EWM, it is necessary to have converted many of the sceptics. The first activity for anyone trying to implement EDM/EWM is to sell the idea. This will have to be done with a mixture of carrots and sticks, bells and whistles. The bells and whistles of a new technology will attract some supporters—in particular, those who are not focussed on day-to-day business results, but are free to dream a little.

Carrots come in the form of opportunities for those who can see that EDM/EWM can help them go further. Sticks come in the form of current problems that can be applied to the backs of those who suffer from them.

Since there is a wide audience of potential customers who may be interested, there is no one easy message for selling EDM/EWM. Within one overall vision, an individually tailored message has to be found for each potential customer. The corresponding leverage, communication mechanism, and timing must be identified.

For an IT specialist, a bells-and-whistles approach may be tried. It may be possible to gain support by providing the opportunity to work with such up-to-date technology. However, this is not the message for the Engineering VP. What carrots and sticks can be found for this position? Reducing the number of engineering changes could be a carrot. What sticks are usually used to beat the engineering VP? (Examples showing engineering in a bad light can probably be found. Perhaps some products have been very late to market, or perhaps exactly the same product has been designed twice.) Could EDM/EWM help to remove them? (If products get to market earlier, customers will provide feedback earlier, and the next version will be even better. At last, engineering will be given credit for its wonderful products.)

The Quality VP may be beaten with sticks such as scrap, rework, and returns. These tend to have secondary costs such as inspecting returned goods, paying late delivery penalties, and carrying excess inventory. If EDM/EWM can be shown to help reduce these, the Quality VP will be grateful.

An even higher level message, referring back to specific corporate objectives, will have to be developed to convince corporate management of the need for EDM/EWM.

1.12 NOW READ ON

To sell the appropriate message at many levels of the company, and in various functions, requires a wide knowledge and understanding. The rest of this book provides information covering a range of business, engineering, implementation, and EDM/EWM system issues. Discussion of the business issues will help the reader to sell EDM/EWM to corporate management, corporate planning, and top management. Understanding of the engineering issues will be necessary for discussions with the Engineering VP and other members of the engineering department. The section on EDM/EWM systems will both help the reader to understand the functionality of such systems and prepare the reader for discussions with IT specialists. The two chapters devoted to implementation issues will help the reader to successfully plan and implement an EDM/EWM system.

2

EDM/EWM Issues for Top Management

2.1 INTRODUCTION TO ENGINEERING DATA AND ENGINEERING WORKFLOW

The term "engineering data" (Exhibit 2.1) includes all data related both to a product and to the processes that are used to design, produce, and support it. Some engineering data (e.g., part geometry) is created within the engineering function, some is created elsewhere (e.g., in field tests). Some of the data (e.g., stress or circuit analysis results) is used within the engineering function, some (e.g., welding instructions) is used elsewhere. Engineering data can be found in many locations and on many media. There is a lot of it (some companies have many hundreds of thousands of documents and many millions of megabytes of data). It comes in many forms (e.g., numeric, graphic, alphanumeric). Some of the users of data will be inside a company, others outside (e.g., suppliers and customers).

The term "engineering workflow" refers to the flow of work through those activities that create or use engineering data. Engineering workflow is not limited to the flow of work through the classical engineering department. It also includes the flow through those other organizations that make use of engineering data. Some of the activities take place inside the company, others outside the company. With this definition, the use of welding instructions on the shop floor is a step in the overall engineering workflow. Similarly, maintenance work is another step. In theory, the engineering workflow starts with initial product specification and ends with product use by the customer. In practice, it is not so simple. Engineering workflow is not a linear process, starting with one well-determined activity and continuing serially through other well-determined activities, until it reaches a

18

ENGINEERING DATA

Product Data and Process Data

Created and Used by the Engineering Functions

Created and Used by Other Functions

In Many Locations, on Many Media

ENGINEERING WORKFLOW

From Specifications Through Shop Floor to Customer

A Complex Nonlinear Process

Characterized by Versions and Changes

EXHIBIT 2.1   Engineering data and engineering workflow

well-determined final activity. Instead, it is a complex process in which some activities run in series and some run in parallel. Sometimes it is only when an activity has been completed and its result is known that the next activity can be chosen. In some cases, the result of the activity will mean that previous activities have to be repeated.

2.2 OBJECTIVES OF EDM/EWM

To be successful in today's highly competitive, global manufacturing environment, a company must be able to supply and support the products that customers require, at the time required by the customer. These requirements put tremendous pressure on the engineering function to improve product quality and reduce lead times. One way to do this is to increase the productivity of individual engineering activities (e.g., through the introduction of CAD). Another way is to improve the coordination between activities (e.g., through concurrent engineering). Both approaches must be followed if maximum results are to be achieved. EDM/EWM supports both approaches.

Currently, most companies would claim to have engineering data and engineering workflow more or less under control. They recognize, though, that this state of affairs may not last for long. There are several different factors that will make both data and workflow more difficult to control (Exhibit 2.2).

CURRENT SITUATION

Just About Under Control

FUTURE SITUATION

High Volumes of Data Easily Produced with Computers

Increased Customization of Products

Increased Role of Suppliers in Engineering Process

Pressure to Improve Engineering Productivity

EXHIBIT 2.2   Engineering data and engineering workflow—the trend

One of these factors is the rapidly increasing amount of data produced by computers. Another is the increasing customization of products. Whereas in the past there may only have been one set of product definition data for a given product, in the future a product may have a different set of product definition data for each customer. There will be a wider range of products, as well as more versions of each product. Information will have to be maintained on the exact product configuration used by each customer and on all of the different assemblies used to make up the product in its different configurations. The increasing tendency of companies to focus on that part of their product where they believe they are the best, and to buy in the rest of the product from external sources, will lead to an increasing flow of data between a company and its suppliers and to an increasingly complex workflow. Market conditions require manufacturers to reduce lead times significantly. Whatever techniques (e.g., concurrent engineering, just-in-time) are used to do this, the effect is the same. There will be even less time available in the future for a process that today can only just be carried out in the time available.

At first sight, it may appear that the need for engineering data management stems from the need to manage the large volumes of data generated by computer-based systems. However, it is actually the business reasons, the needs to improve engineering productivity and to respond more flexibly to customers, that have become the driving force to achieve better engineering data and engineering workflow management. EDM/EWM oversees the creation and use of engineering information throughout a product's life. As Exhibit 2.3 shows, it is by improving the use, quality, and flow of engineering data and by supporting new engineering techniques that EDM/EWM will make it possible to reduce lead times and product costs, and improve competitiveness, market share, and revenues.

IMPROVE
The Use, Quality, and Flow
of
Engineering Data

and

SUPPORT
New Engineering Techniques

to

REDUCE
Lead Times and Products Costs

and

IMPROVE
Competitiveness, Market Share, and Revenues

EXHIBIT 2.3   The objectives of EDM/EWM

2.3 KEY ISSUES FOR TOP MANAGEMENT

Exhibit 2.4 shows seven key engineering data management/engineering workflow management (EDM/EWM) issues for top management.

Management needs to get engineering data under control. Engineering data lies at the heart of the engineering function in today's computerized, information-based environment, yet few companies have it under tight control. Data is scattered in many locations. It is on paper, microfilm, aperture cards, and computers. Most engineers do not fully trust the system and, to protect themselves, keep several copies of data and try to stop other people accessing their data. Often, several groups of users will have different versions of what should be the same data. Each will claim to be the legitimate owner of the data and, hence, the owner of the correct version. As companies invest in more and more engineering workstations, thus creating a truly distributed computing environment, it becomes even more difficult to maintain control of the company's engineering computers and to maintain control of engineering information, a valuable company resource. It becomes more difficult for companies, that must comply with legal requirements on traceability, to maintain audit trails so that they can track back to the source of any product problems. It becomes increasingly difficult for companies to know which is the "master" version of data—is it a computer-based model or a drawing,

Control of Engineering Data

Quality of Engineering Data

Reuse of Engineering Data

Security of Engineering Data

Availability of Engineering Data

Control of Engineering Workflow

Support for New Engineering Strategies

EXHIBIT 2.4   Key EDM/EWM issuses for top management

is it the scanned image of a document, or the document itself? As more data is shared with suppliers, similar questions arise as to which company has the master version. Procedures have to be defined to show how both the company that has the master version and the company that does not have the master version can make changes.

Management needs to make sure that the engineering data in use is of a high quality. Without reliable, timely, and accurate information, managers and users cannot work efficiently. It is not easy to run quality checks on data moving invisibly around networks. Instead, quality has to be built in. This can only be done through the right procedures and a company culture that penalizes poor quality work. Error creation and propagation must be prevented. It is only too easy for a user to introduce an error into data. Once the error is in, though, it can be difficult to find, and it can be even more difficult to remove its effects.

Management needs to make sure that engineering data is reused and that it is allowed to evolve. There is no point in continually reinventing the wheel, yet designs need to be revised so that they continue to meet market requirements. Reuse of information was supposed to have been one of the major advantages of using CAD/CAM, but in practice there has been less reuse than expected. The reason has been the difficulty for users, even if they are aware of suitable existing data, to actually find this data. It may be somewhere on the CAD/CAM system or somewhere in a drawing store. Few users are prepared to spend hours or even days hunting for existing data. Instead, they will take a clean sheet or screen and design a new part.

Management needs to ensure that its engineering data management strategy addresses both existing data and data that will be created in the future. Engineering data management is not a green field activity and should not be treated as one. Most companies have a vast amount of information tied up in existing data. One of the major challenges to management when developing an engineering data management strategy is to marry the ability to meet current and future needs effectively with the capability to reuse existing data.

Management needs to make sure that engineering data is secure. There are many facets to the security needs for engineering data. They range from the individual user worried that a colleague might unintentionally overwrite a file to the company that worries that access rights granted to trusted suppliers might somehow be discovered by unscrupulous competitors. Major multinationals transmitting design information by satellite between sites on different continents wonder how secure their data is. Similarly, defense organizations, while promoting increased use of digital data and electronic data repositories as a means of reducing project costs, are aware that major effort will be required to maintain data security. Management must ensure that confidential and proprietary information is protected from unauthorized access.

Management will have to make data available to users when they need it; otherwise, valuable time will be lost. If management really wants to reduce lead times, it will have to cut the waste out of the engineering process. Administrative paper shuffling will have to be abolished. Just-in-time (JIT) techniques, successfully applied on the shop floor, will have to be applied in engineering functions. Engineering data must flow smoothly through the organization. Old habits of spending hours or days looking for data will have to disappear. The data made available will, of course, have to be the right data; otherwise, more time will be lost until the correct version is found. Within this context, a user may not only be someone working a few yards away from where the data is stored. It could be someone in another organization on the other side of the world, in which case opportunities abound for losing many hours of work.

Management needs to take control of the engineering workflow. Resources (people and systems) must be better managed. In many companies, few people can really describe the engineering workflow, and even fewer know why it takes this shape. In most cases, the flow results not from a reasoned design, but from a long series of minor reorganizations that resulted from changes in departmental structures, product characteristics, and human resources. Managers should carry out a review of the current workflow from initial product specification down to customer support. This will provide them with a solid base from which improvements can be made. Activities that do not add value should be removed, and activities that were previously carried out in serial should, wherever possible, be run in parallel. The management of changes, modifications, versions, and variants will have to be improved. Unless configuration management is improved, the rapidly increasing number of changes and of different versions of products will lead to configuration control getting out of hand.

Management must make sure that EDM/EWM fits into the overall engineering strategy. Improved management of engineering data and the engineering workflow will provide an integrated framework within which management will be able to adopt and support strategies to reduce lead times and cut product costs. On-time availability of data reduces lead times.

Concurrent engineering techniques, implying parallel activities, will not work efficiently unless the workflow is under control and data ownership and usage rights have been defined. Design for manufacture and similar techniques that aim to involve manufacturing staff early on in the design process can only benefit from a clearly defined workflow and the certainty that all parties are working with the latest version of the data.

2.4 EDM/EWM SYSTEMS

EDM/EWM systems are computer-based systems that help manage engineering data (Exhibit 2.5) and the engineering process (Exhibit 2.6). They are used to manage product and process information. Some of this information, such as 3-D CAD models and finite element analysis models, will be generated and used by the engineering function. Some of it, such as parts lists and programs for numerically controlled (NC) machines, will be generated by the engineering function for use

Functional specifications	Process plans
Existing designs	Bills of materials
Parts classifications	Quality assurance data
Design specifications	Version management data
Scanned-in drawings	Change control data
Assembly drawings	Maintenance information
Engineering drawings	Technical publications
Analytic models	Costing data
Simulation results	Test results
Parts lists	Status reports
Computer programs	Purchasing information
Tool and fixture designs	Shop floor instructions
NC programs	Schedules
Spares information	Field failure reports
Standard times and costs	Machine libraries
Test data files	As-designed configurations
Mounting instructions	Flowcharts

EXHIBIT 2.5   The scope of engineering data

Conceptual design	Styling
Simulation	Kinematic analysis
Detailed design	Schematic design
Quality assurance	Technical publications
Configuration management	Project management
Tool and fixture design	NC programming
Process planning	Production planning
Plant layout	Material handling
Reliability	Kitting
Testing	Sales
Maintenance	Purchasing
Marketing	Costing
Packaging design	Quotation preparation
Vendor performance	Assembly
Forecasting	Sales order processing
Drafting	Value analysis

EXHIBIT 2.6   Processes using engineering data

by other parts of the company. Some, such as maintenance information, will be generated by the engineering function for use outside the company. Some of the product and process information will be generated and used outside the engineering function (e.g., in inventory control, in FMS scheduling). Some of it will be generated outside the engineering function (e.g. in quality control, in field testing, from customer complaints) and used by the engineering function.

Many computer-based systems create engineering data. A CAD system is an excellent example. Such systems typically focus on activity-specific functions. They create data, but neglect data management functions such as data definition, structuring, organization, storage, retrieval, archival, protection, distribution, and tracking.

EDM/EWM systems are used in the management of the activities associated with the development and use of the information mentioned above. These processes start with the specification of a product and include geometry definition, analysis, manufacturing engineering, shop floor activities, and product support. Workflow within the overall engineering process is very complex. Many people and many organizational entities are involved. Some activities overlap and others run in parallel. Sometimes iteration is needed. At other times, it may be necessary to repeat a sequence of activities. EDM/EWM systems manage the flow of work through these activities. They provide support to the many activities of engineering such as the design process, the sign-off process, the sharing of data between multiple users, the tracking of engineering change orders, the management of design alternatives, and the control of product configurations.

EDM/EWM systems address both information and workflow issues. As such, they are true integration tools. In particular, within the engineering environment, they are central to the integration of previously separated systems such as computer-aided engineering (CAE), computer-aided design/computer-aided manufacturing (CAD/CAM), electronic publishing (EP), configuration management, process planning, document scanning and project management. Their use will result in improved quality, flow, and use of information related to the engineering process. This will help companies meet the growing demands of an ever more competitive business environment.

2.5 THE IMPORTANCE OF EDM/EWM SYSTEMS

As companies come under increasing competitive pressure, they are required to reduce lead times, reduce costs, and increase quality. The engineering function of the company is a prime target for action to achieve these targets. It is the function that has the major influence on the amount of time that elapses between the moment that the decision is taken to introduce a new product and the time taken to get the product to market. Similarly, it has a major influence on the cost of a product. Although only some 10 percent of total product cost is actually incurred during design and engineering, some 75 percent of costs are defined during these activities. Engineering is the function that is primarily responsible for defining the cost of a product. If the product it specifies is too expensive for the customer, the best efforts of the other functions will be wasted. Since the quality of the product delivered to the customer is in many ways a function of the quality of the product defined by engineering, the engineering function can play a major role in improving product quality. Engineering is an upstream function. If it makes mistakes, the downstream functions, such as shop floor operations, will suffer. A change made before a design is released is relatively cheap to correct. A change made once a product is in production may be hundreds or even thousands of times more costly to correct. By improving the control of engineering information and activities, EDM/EWM systems will help reduce lead times, reduce costs, and improve quality.

Reduced lead times open up new market opportunities and improve profits. They also reduce market risk by reducing the time between product specification and product delivery. The sooner customers use a product, the sooner their feedback can be incorporated in a new, improved version. If quality is improved, not only will customers be pleased, but there will be a reduction in scrap, recall, and rework. Corresponding administrative activities and their costs will be reduced.

EDM/EWM will improve engineering productivity. Engineering managers will know the exact design status. They will be able to assign resources better and release designs faster and with more confidence. Design engineers will know which parts are available and which procedures should be followed when designing new parts. Manufacturing engineers will be able to see how similar parts have been made previously. Everyone will be able to rapidly identify approval mechanisms.

In many high-tech companies, the engineering function is expensive to run, as well as being a major customer for investment. Any increase in its internal efficiency will have a positive effect. As long as engineering remains an apparently artisanal and uncontrollable process, management will find it difficult to achieve the expected improvements in engineering activities. EDM/EWM systems help in attaining the goal of a manageable engineering process that makes the best possible use of engineering information. Top management involvement will be necessary to meet this important goal. Before looking at some of the reasons why such involvement is so important, some of the practical problems that occur in the absence of EDM/EWM will be described.

2.6 THE ENGINEERING ENVIRONMENT WITHOUT EDM/EWM

In the typical company with large numbers of people involved in the engineering and associated functions, many problems arise from the lack of control over engineering data and the engineering process (Exhibit 2.7). People are sometimes unable to find the information they need. If they can find the information, it may not correspond to the actual state of the product. For example, a facility drawing may not correspond to the physical facility layout.

Developers are unable to rapidly access a particular design among the mass of existing designs. To find specific information, they may have to search through tens of pages of listings. They lose valuable time. Studies show that design engineers spend up to 80 percent of their time on administrative and information retrieval activities. They develop new designs that may be almost identical to existing designs, with the result that unnecessary additional costs are generated as the new designs are taken through the various activities necessary for manufacture and then supported during use.

As more and more data is generated on computer and other electronic systems, it becomes more and more difficult, with manual control and management procedures introduced when data volumes were 20 or even 50 times lower, to track the location of data, to prevent unauthorized access, and to maintain up-to-date product configurations. Large companies hold many hundred TB (terabytes) of

Data is lost. Data is not secure.

Data cannot be found. Time is wasted searching for data.

Data does not correspond to the product.

Unnecessary copies of data are maintained. Data is not shared.

Copies of data conflict. Data ownership is not clear.

Islands of automation abound.

Inter-functional communication of data is slow.

Transfer of data on paper is slow.

Review and approval procedures take too long.

Engineering change management is too slow.

Engineering changes are not controlled or coordinated.

Project managers lack up-to-date status.

Projects overrun. Project costs are unnecessarily high.

Lead times are unnecessarily long.

Versions are not properly managed.

Configuration management breaks down.

Configuration information does not correspond to reality.

Technical manual development is labor-intensive and slow.

Technical manuals are out of date.

Maintenance information is inaccurate.

Traceability cannot be maintained.

Computer programs are not documented properly.

EXHIBIT 2.7   The engineering environment without EDM/EWM systems

data (Exhibit 2.8). Companies hold thousands or even millions of drawings. 3-D CAD part models may take up several megabytes. One company calculated that it needed 250,000 pages to describe a new product and that, on average, each of these was reproduced 30 times. This can be thought of as many tons of paper, or many gigabytes of data.

Data entry is poorly controlled. Data is lost and cannot be retrieved. It is re-created and errors are introduced. The wrong product goes to a customer. Product configuration data is not up to date. When a defective part is found in the

KB	1 000 bytes
MB	1 000 000 bytes
GB	1 000 000 000 bytes
TB	1 000 000 000 000 bytes

EXHIBIT 2.8   Kilo, Mega, Giga, Tera

field, many more products than necessary have to be recalled. Design history is not maintained, so it is impossible to draw on previous experience.

Due to gaps between incompatible application programs, data is transferred manually, and errors occur. They have to be corrected, and their correction has to be managed. This costs time and money. An error slips through and is not discovered for several weeks. Correcting it leads to months of delay. Part descriptions and bills of materials developed on a CAD system may need to be manually transferred to an MRP 2 system in a computer that is not linked to the CAD system. The manufacturing bill of materials may be different from the engineering bill of materials. The two systems may be the responsibility of different departments or organizations. The change processes in the two organizations may be different and out of step. At a particular time, a given change may have been made in one system, but not in the other. As a result, not all users will have immediate access to the most up-to-date information. Thousands of dollars can be wasted if manufacturing works on the wrong version of a design for several weeks.

Several copies of the information describing the same part are maintained. Nobody knows which is the master copy. When a change is needed, not all copies are changed and not all downstream functions are alerted. Old, unwanted revisions of parts are machined, while the new, required versions are ignored. When there is no agreed master version of a particular item of information, and no agreed owner, all users of the information will behave as if they were the owners. Each user will define the item to suit their particular requirements. All the definitions may be different. Such wasted effort leads to confusion when information is transferred between users.

Attempts to work closely with a supplier in the design phase are hindered because it takes several days (a month in the case of one aircraft manufacturer) to transfer a paper drawing from one company to another. Increasingly, companies expect suppliers to design and produce complete assemblies containing many parts. Unless the company's after-sales engineers have access to information on individual parts of the supplier's assembly, problems arise in the field.

Project and resource management tools are not linked to the design information. Unintended overlap in data and workflow occurs, wasting time and money. At the same time, any attempt to save time by running the various phases of a project in parallel leads to chaos. As a result, the phases are run in serial, lengthening project cycles. Rules and procedures are difficult to enforce. Design rules can be ignored. Project planning exercises cannot draw on real data from the past, but are based on overly optimistic estimates. Project managers find it difficult to keep up to date with the exact progress of work. As a result, they are unable to address slippage and other problems as soon as these occur.

Engineering changes are poorly coordinated, with the result that unnecessary changes may be introduced. Design cycles are longer than necessary, and un-released versions of data are acquired by manufacturing, causing confusion and waste. The time taken for raising, approving, and implementing changes becomes much longer than necessary. The change process may take days, weeks, or even months, whereas the actual processing time may be only minutes or hours. In large companies, it costs thousands of dollars to process an engineering change.

Engineering change control systems are often bureaucratic, paper-intensive, com-plex, and slow. A central engineering services group may have the responsibility but not the tools to push the changes through as quickly as possible. Many departments may be involved (one manufacturer found that, depending on the change, up to 16 departments were involved). As a result, it may take several months and 50 or more different documents to get a proposed change approved and incorporated into the product design. During this time, the product will continue to be produced with an unwanted design. Even when a change has been agreed and announced, many months may go by before the corresponding documentation gets to the field.

As the management and change process appears as an inefficient and time-con-suming overhead, some people will avoid it. For some information, there may even be no formal change control process. Minor modifications to products and draw-ings will not be signalled. Components will be substituted in end products without corresponding changes being made to test routines. People will fail to maintain the trace of the exact ingredients in ever smaller batches of products. Nobody will notice until something goes wrong or another change has to be made. Then, unnecessary effort will be needed to find out where the problem comes from. Additional support staff will be employed to try to prevent further problems.

Informal communications are developed between departments to cope with the lack of suitable formal communications. Few records will be kept of this type of transfer, and, in the absence of a particular individual, it may be impossible to find any trace of important information.

People wait hours, even days, for a given piece of information. The person who should sign off a design is called away for a few hours, and work is held up because nobody knows who else has the authority to sign off. When people do receive informa-tion they are not sure if they have received the correct version. Sometimes they just want to make a simple request for information from a system that "belongs" to some-one else, and find they have to wait several days to get it.

Configuration control breaks down. Configuration documentation no longer corresponds to the actual product. Unexplained differences appear between as-designed, as-planned, and as-built bills of materials. Increased scrap, rework, and stock result. Incomplete products are assembled and delivered. Field problems are difficult to resolve, and inefficiencies occur in spare parts management. New versions of computer programs are introduced without sufficient care being taken to ensure that, for example, data created with earlier versions is still usable. Programs are insufficiently documented. Program modifications are made without appropriate change control.

Technical manuals become outdated, yet are not updated. Logistics support data gets out of control. Inadequately documented configurations become difficult to maintain. Spares replenishment becomes inaccurate, and customers have to immobilize products while efforts are made to identify correct replacement parts. When the right part arrives, the right handling equipment and maintenance tools are not in place.

There is a conflict between the central electronic data processing (EDP) staff and the computer system support staff in the engineering function. The EDP staff are deeply involved in F&A work, and they do not invest enough time in supporting the engineering function. They are often more interested in the theoretical aspects of EDP and new EDP technologies than in its practical use for engineering activities.

New product introductions in a whole range of manufacturing industries (Exhibit 2.9) are delayed for a myriad of apparently random and minor, but cumulatively significant, reasons. Product quality is erratic despite vast investments in engineering and manufacturing technology and in quality programs. Lead times seem to remain the same in spite of all the new investment. Overall, the costs associated with engineering and development rise rather than fall.

Aerospace	Iron and steel
Architecture	Machine tool
Automotive	Machinery
Beverage	Mechanical engineering
Chemicals	Offshore engineering
Computer	Pharmaceutical
Construction	Pulp and paper
Electrical	Shipbuilding
Electronics	Software engineering
Energy	Telecommuncations
Food	Textiles and clothing
Industrial equipment	Toys

EXHIBIT 2.9   Industries that can benefit from EDM/EWM

2.7 TOP MANAGEMENT INVOLVEMENT WITH EDM/EWM

Before looking in more detail at the technical solutions that EDM/EWM systems can offer, it is useful to outline some of the strategic reasons for the importance of EDM/EWM and the corresponding need for top management involvement (Exhibit 2.10).

Unless top management is involved, there is a danger that EDM/EWM will become a victim of the general lack of vision and coordination that many companies exhibit towards the use of computers. In the past, EDP managers, often reporting to finance & administration (F&A) managers, would develop an EDP plan for their use of computers. This generally addressed the use of computers in the finance function, but rarely considered the use of computers and telecommunications in other functions such as marketing, engineering, and manufacturing. As information technology (IT) has become a tool with which a company can gain a competitive advantage, it has become an important component of corporate strategy. During the corporate strategy development process, account should be taken of the possibilities that IT can offer in helping to meet the corporate goal of providing optimum products and services to customers. In many companies, IT is no longer seen as an appendage to F&A, and the IT strategy is given the same weight as the marketing, engineering, and manufacturing strategies. It becomes a critical component of the corporate strategy, as it also underlies the functional strategies. In companies that limit their vision of IT to the use of EDP in F&A, EDM/EWM has little chance of success. EDM/EWM must be seen as part of the

---

Engineering information is an important company resource.

Engineering is a key function.

Implementation has major organizational implications.

Information and workflow are cross-functional.

EDM/EWM must link with the overall IT strategy.

EDM/EWM must support the engineering philosophy.

Successful implementation takes time and money.

EXHIBIT 2.10   Reasons for top management involvement with EDM/EWM

overall IT strategy. It cannot take its rightful place if top management is unaware of its existence and potential.

Typically, in spite of the importance of computer-aided tools in engineering, companies will have an engineering (or an R&D) strategy, but not a strategy for the use of computers and telecommunications in engineering. Instead, computer and communication systems will be acquired by engineering departments, groups, and projects to meet the demands of a particular project, but with no overall objective behind their acquisition.

In today's high-tech companies, the engineering function could not function without computer systems and a wide range of application programs and communications. In practice, the engineering systems and applications are often discrete and poorly integrated. In many ways, they create as many problems as they solve. EDM/EWM systems are the tool that can link them together and make sure that they fit together well as part of the overall process.

EDM/EWM systems support the strategies and philosophies often currently adopted by companies, be they total quality management (TQM), concurrent engineering, or a focus on excellence. A fundamental component of TQM is the control of the functional processes. EDM/EWM offers the possibility to increase control of the complex and difficult-to-manage engineering process.

In the traditional engineering process, products were developed in a series of separate steps, starting with design and engineering. As a second step, manufacturing engineers tried to work out the best way to produce parts. Once this had been done, the purchasing department would look for the best sources of parts. Eventually, an attempt would be made to manufacture the product. Design errors and incompatibilities would then be seen, and much time and money would be spent on making the necessary changes. Concurrent engineering and simultaneous engineering, with their emphasis on product teams made up of individuals from different departments and even different companies, and parallel working on processes that were previously carried out in series aim to overcome the disadvantages of the traditional method. They are only feasible in a well-controlled engineering environment, with a high availability of engineering information and rapid communication of information between individuals. EDM/EWM provides a solution. Design for manufacture techniques require up-front involvement of manufacturing personnel in the design process. EDM/EWM helps different groups share information and communicate effectively. Many manufacturing companies are trying to focus on the aspects of their products or services where they excel or where their value-added is highest. This implies partnerships or contractor relationships with other companies that will work on the other parts of the engineering/manufacturing process. These relationships in turn require efficient information transfer and management between the partners, as well as close control over projects involving several partners. Again, EDM/EWM systems help meet these requirements. As well as supporting strategies, EDM/EWM systems can be a source of new opportunities. If a company can radically improve its engineering performance through the use of EDM/EWM systems, it may be able to significantly change its market position.

A further factor pointing to top management involvement is the general trend towards operations becoming more information-intensive, and the corresponding push to use information-driven processes to support overall business activities. Another is the increased computerization of engineering activities and the resulting sharp growth in the volume and availability of digital engineering data. Together, these will make engineering information an important company asset, requiring top management attention. This will lead to an awareness of the importance of engineering data management and the advantages that can be achieved through improved control of the engineering process. Information security, privacy, and backup will become key issues.

There are also organizational reasons behind the need for top management to become involved with EDM/EWM. Engineering data is used and reused by many of the organizational entities within the company. The approach to its overall management has to be cross-functional. Attempts made by individual departments (such as design engineering) to create or impose order will be frustrated by other departments (such as manufacturing engineering) that may feel threatened by such moves or feel it necessary to assert their independence. If this were to happen, the gains expected from an integrated approach would fail to appear.

The introduction of EDM/EWM is a long-term process. For medium and large companies it may take anywhere from two to seven years. The combination of cross-functional and long-term project characteristics implies that the introduction of EDM/EWM will be difficult to cost-justify. Cost-justification techniques that only take account of short-term, task-oriented benefits that can be indisputably traced back only to the introduction of EDM/EWM will fail. The reduction of personnel in the "drawing store" is easy to measure, but will not pay for the introduction of EDM/EWM. The more important benefits of EDM/EWM are cross-functional and long term, and relate to time-to-market and product quality. Although these gains are easy to calculate, it can be difficult to measure the component that is due to EDM/EWM.

Another important point to bear in mind is that although, at first glance, EDM/EWM looks very much like a purely technological solution to the problem of data management, it is not. It implies major organizational effort. Experience (from the introduction of CAD/CAM, for example) has shown that a "technological solution" is rarely successful unless it is put in place alongside the corresponding "organizational solution." Many companies bought a CAD/CAM system and tried to use it without taking the appropriate organizational actions. The net result was that the CAD/CAM system had little effect on the engineering process. However, in companies where appropriate organizational actions were taken, use of CAD/CAM has lead to significant benefits. Typical actions include setting clear targets for system use, showing management how to manage the system, training people to use the system, training a system support team, defining system use procedures, modifying workflows, and defining and using a suitable design methodology.

When EDM/EWM is introduced, there may be great pressure to install it and get it working as cheaply as possible. This can be expected to lead to minimum, or zero, investment in the organizational aspects and, a few years later, the realization that the system is not fulfilling its initial promise. To avoid this unfortunate result, top management must be involved in the early stages of introducing EDM/EWM and must make it clear that the initial objective is not to select the solution that appears quickest and cheapest. Instead, top management must ensure that the decision takes full account of the company's EDM/EWM objectives, the role of EDM/EWM in the future, and the organizational actions that must be taken for an apparently technological solution to be used successfully. A key part of this decision will be the cost justification of EDM/EWM.

2.8 THE NEED FOR EDM/EWM

The detailed technical needs for EDM/EWM systems are described at length in the next chapter. Briefly, they can be divided into two classes. Some positively impact engineering operations, and some reduce the problems that occur in the engineering environment. In practice, both classes of reasons are sources of opportunity. Taken from the positive viewpoint, EDM/EWM systems offer the potential for better use of resources, better access to information, better reuse of design information (since this will be under better control), better control of engineering changes, a reduction in design cost (since it will be easier to be aware of real costs during the engineering phase), a reduction in lead times, and improved security of engineering information.

EDM/EWM systems help companies to improve their competitive edge. They help improve the productivity of the engineering process. They allow companies to be more flexible in their manufacturing. They help companies improve the quality of their products, and they allow these companies to be more adaptable to market requirements.

Looked at from the other point of view, EDM/EWM systems can be thought of as systems that solve some currently existing problems. There is a rapidly increasing amount of data available in the engineering environment. Some of this is on electronic media (disks, tapes, and cassettes), but a lot is still on paper and other traditional media. (Few companies have even 10 percent of their engineering information on electronic media.) There are different types of data. There is numeric data (such as part geometry in a CAD system), text information (such as specifications and technical publications), graphic information (such as engineering drawings, photographs, and films) and, increasingly, voice data. Without EDM/EWM systems, this information cannot be managed efficiently.

As companies increase their use of engineering drawing scanning systems, they need EDM/EWM systems to manage the large volumes of data that result and to control access and use of scanned data. In the CAD/CAM area, EDM/EWM systems can help solve problems arising from the vast number of part files and versions that can so easily be produced. If a company uses several different CAD systems and has to transfer data to suppliers with other systems, the data management process can easily get out of hand. EDM/EWM systems can help regain control.

Experience with the introduction of CAD/CAM provides some useful hints for the introduction of CAD/CAM. Many companies, particularly those working in general mechanical engineering, find it difficult to prove any real gains from the use of CAD/CAM. This is not to say that CAD/CAM is never useful. There are application areas for CAD/CAM, such as electronic design and the design of molds for plastic parts, that are clearly very successful. In many cases, a specific reason for this can be found. Electronic design is primarily a two-dimensional application, unlike general mechanical engineering, which has to handle three-dimensional parts. Electronic design is therefore "simpler." The design of plastic molds is, of course, a three-dimensional application. However, it often involves working with complex-surface geometry. With manual techniques, this process is very lengthy. CAD/CAM significantly reduces the time required.

In both electronic design and mold design, successful companies have a high level of integration between different parts of the engineering process. In electronic design applications, simulation, design, layout, and test systems are closely linked, and the data in the system is used effectively. On the other hand, in typical mechanical engineering companies, which have a much wider range of processes to handle, the same is not true. Instead, systems usually appear as separate "Islands of Automation," and a lot of time and effort is wasted in managing information in such an environment. It is difficult to control the exchange of data between different systems in the absence of well-defined and properly enforced procedures. A related problem that arises is that the same data is defined and modified in different parts of the organization, with the result that there is no agreement as to the correct version.

The problems that arise from lack of data management can also be looked at from a level higher than that of the data itself. As a result of poor data management, engineers may waste time finding out which version of a part they should work with, or discovering which is the latest version of an existing part. Designs will take longer than expected, because so much time is spent looking for data and checking that the right version is being used. Sometimes the wrong information will be passed on to other people in the organization with the result that their time is wasted, and, in some cases, parts are made that will eventually have to be scrapped. A key area here is the interface between the engineering organization and the manufacturing organization. Unless the data release process is under control, time and effort will be wasted.

2.9 EDM/EWM IN THE 1990s

One of the main factors behind the rapidly emerging need for EDM/EWM is the growth in data produced by CAD/CAM systems. Traditionally, these systems have been file-oriented. While data volumes were low, such a solution was acceptable, but as data volumes mushroomed in the late 1980s, another solution became necessary. Those companies that became aware of their need to manage engineering data looked to the data base systems then on the market (i.e., those used for commercial and financial applications) for a solution. As will be seen later, without major modification, these are not suitable for the engineering environment. It was not until the late 1980s that a few companies started to develop data management systems specifically for the engineering market.

A second factor behind the growing interest in EDM/EWM systems has been the CALS program of the U.S. Department of Defense. As will be seen in Chapter 8, one of its major thrusts has been towards the digitalization of paper drawings. Once a company starts electronic scanning of its drawings, data volumes grow very quickly, and a data management system is needed.

Another important factor is the need to get control of the engineering process and to reduce engineering cycle times. A prerequisite for this is better understanding and management of the engineering workflow.

By the end of the 1980s, the need for EDM/EWM systems had been identified. It was already possible to see that EDM/EWM would become an important part of the computer-aided engineering (CAE) environment. However, at that time, very few companies were actually using EDM/EWM systems. The new decade began with only a few thousand of the potential 3 to 5 million users worldwide equipped with EDM/EWM systems. From that small base, rapid growth in the number of systems installed was expected from 1991 to 1995.

2.10 EDM/EWM "DOS AND DON'TS" FOR TOP MANAGEMENT

Do remember that introduction of EDM/EWM is lengthy, costly, and cross-functional. Don't look for a quick, cheap solution for one department.

Do run EDM/EWM as a cross-functional, high-level activity reporting directly to top management. Don't treat it as a local issue that can be solved by a few programmers and data analysts.

Do take the time to understand the many interrelated issues involved in EDM/EWM. Don't try to split off and solve one issue before attaining an overall understanding.

Do start by trying to understand the business objectives. Don't start by trying to model all the data flows in the engineering department.

Do consider the many forms of data. Don't only address CAD data, or paper drawings, or alphanumeric documents.

Do take the opportunity to improve the engineering workflow. Don't automate activities that add no value.

Do take account of customers. Don't only focus on internal activities.

Do take account of organizational and cultural issues. Don't believe that purchasing an EDM/EWM system will automatically lead to successful use of EDM/EWM.

3

Engineering Workflow Management

3.1 THE IMPORTANCE OF EWM

The term "engineering workflow" refers to the flow of work through those activities that create or use engineering data. It is as important for companies to manage engineering workflow as it is for them to manage engineering data. It is only too easy for companies to get so immersed in the many highly visible problems of managing data that they neglect the workflow. If they do this, though, they will probably not achieve their EDM goals, such as helping to reduce product development lead times and engineering change management costs. In companies that still have traditional, sequential engineering workflows, the inefficiencies of the workflow can nullify the benefits of improved data management.

The product that the customer will eventually receive is designed and manufactured by the activities of the workflow. This means that the quality and cost of the product are functions of the engineering workflow. The elapsed time between the first idea for a product and the moment that the first customer receives the product depends on the efficiency of the engineering workflow. Since the engineering workflow affects product cost and quality, as well as manufacturing and overall lead times, companies faced by competitors producing higher quality products faster and at lower cost need to improve their engineering workflow. Before being able to do this, they must first understand the workflow, and bring it under control. Unless they take control of it, they cannot hope to improve it and, as a result, improve customer service.

It is not difficult for companies with a traditional engineering workflow made up of sequential independent steps (conceptual design to preliminary design to analysis to detailed design to process planning to purchasing to production planning to production to quality control to field tests) to understand where the problems lie. Disjunctures, superfluous steps, and inefficient activities in the workflow all contribute to unnecessarily extending lead times, increasing costs, and reducing quality. With the traditional workflow, it is impossible to accurately forecast costs and lead times at the beginning of the process because nobody knows what will actually happen. In any given step of the workflow, people do not have all the necessary knowledge and experience, so they make assumptions and get something wrong. Later in the sequence, a correction has to be made. The process loops back and time is lost. In some cases, the change process takes up as much time as initial development. The change process is so complex that bureaucratic approval procedures have to be added between steps to govern changes. In one company, more than 20 managers were having to sign off individual steps and changes. This not only wasted the manager's time, but also lengthened the development time. In many industries, time to market is the key parameter. An overrun of six months will cause a big drop in expected profits and have a very negative effect on customers.

Most companies are organized to manage individual steps in the workflow, but the overall engineering workflow is not managed. Individual activities are, at best, managed on a departmental or functional basis, and tend to take account of neither the overall flow, nor the detailed needs of other groups. Often, no one individual has the responsibility for the overall workflow. The problems are neither understood in detail, nor addressed from an overall point of view. As a result, they are not solved. The overall workflow remains inefficient, wasteful, and out of control. In this state it cannot be optimized.

When overall engineering workflow is brought under control, lead times are reduced, quality goes up, and costs go down. Once the workflow is controlled, a good understanding of final product costs is achieved earlier in the engineering process. Accurate estimates of lead times can be made.

3.2 THE LINK BETWEEN EDM AND EWM

All the activities along the engineering workflow create and/or use engineering data. The workflow exists to provide the engineering data necessary to produce the product. Without engineering data, there would be no need for the engineering workflow.

Each step, or activity, in the workflow has its own information needs, information input, and information output. Within an activity, people use information. If information is not available, it may not be possible to complete the activity. Often, the end of an activity is characterized by information being prepared, signed off, and released. Between activities, information is transferred. When an iteration or change occurs in the workflow, corresponding information is produced. Information flow has to be synchronized with workflow so that the information is available when and where it is needed.

Within individual engineering activities, the percentage of time individuals spend looking for or transferring information is high. For many otherwise productive individuals, it may be 30 percent or even 50 percent. As time may also be taken up by management tasks, the time actually spent on the functional activity may only be about 30 percent. In addition, in many companies there are technical "liaison" staff who spend 100 percent of their time looking for information that should have been transmitted by other departments.

Engineering data is created to be used by someone else. Presumably, the creator of information knows who it is being created for. Once created, it should be moved on, it should flow to the activity that is going to use it. Since engineering data and engineering workflow are so closely linked, it is not possible to control one without becoming involved with the other. Implementation of major changes in the way that data is managed requires a reappraisal of the way that the engineering workflow is managed.

When automated engineering data management is introduced, major steps are taken, either deliberately or by default, towards automating the workflow. If the workflow is not improved before the information flow is automated, the existing, inefficient workflow and information flow will be automated.

3.3 ENGINEERING WORKFLOW

Engineering workflow can be characterized by a variety of parameters, such as product/market features, activities, costs, time, resources, number and frequency of iterations, number and frequency of changes, and extent of parallel activities. Exhibit 3.1 shows examples of some of the activities of the engineering workflow. Traditionally, companies have thought of these activities as taking place in series rather than in parallel.

The activities followed by a particular company will depend greatly on the type of product and the company's position in the value chain. Within process industries, there will not be the same workflow for a producer of fine chemicals as for a producer of bulk chemicals. Similarly, in discrete manufacturing industries, the workflow will be different for producers of airplanes, machine tools, and durable consumer goods. Even between companies in the same sector, workflow will be different. Differences will be due to different customer bases, product life cycles, production runs, new product introduction rates, and special regulations, as well as due to company size and organizational structure. Workflow is company-specific. There is no easy solution available allowing a company to improve its workflow without first understanding all the details of its workflow.

The approach to changes of both product and process specifications differs from one company to another. Changes cost money, waste time, require iterations in the workflow, and require extra management effort. When design changes occur late in the workflow, they can be extremely expensive, impacting the product,

Marketing	Release to manufacturing
Opportunity assessment	Production planning
Mission definition	Sales
Project management	Purchasing
Specification	Order processing
Product definition	Manufacturing
Conceptual design	Assembly
Feasibility	Inspection
Preliminary design	Shipping
Analysis	Installation
Prototype building	Support
Prototype testing	Use
Detailed product design	Modification
Documentation	Maintenance
Process design	Obsolescence

EXHIBIT 3.1   Some activities of the traditional engineering workflow

production equipment, and launch date. In the worst case, the customer will receive a defective product.

It is important to understand where a company defines costs in the workflow, and where costs are actually incurred. In the same way, it is necessary to understand where quality and time cycles are defined, and where they become reality. Various studies have been made of the distribution of costs in the traditional engineering workflow. Typically, they show some 60 to 75 percent of costs defined (and 1 to 5 percent of costs incurred) during conceptual design, and 85 to 95 percent of costs defined (and 10 to 15 percent incurred) before release to manufacture. The incurred cost of engineering changes during conceptual design is negligible compared to that of changes after release. Fixing a design problem when a product is in the field is often thousands of times more expensive than preventing it during initial design. It would seem to be sensible to spend more effort on getting the conceptual design right than on doing it quickly and making expensive corrections later. Simultaneous engineering, described below, follows this approach.

The above figures do not take account of product use by the customer. Some products require a great deal of support and maintenance. The price of maintenance and spare parts for some products exceeds their purchase price. For example, airlines typically spend two or three times the purchase price of an engine on spare parts for the engine. Support and maintenance activities can significantly change the distribution of life cycle costs, often reducing even further the percentage of costs incurred during the early development phases. They should be considered as part of the engineering workflow and taken into account when deciding how to improve the workflow and where to assign resources.

3.4 AN INEFFICIENT WORKFLOW

As Exhibit 3.2 shows, the engineering workflow is inefficient. It is expensive, slow and bureaucratic and, invariably, takes longer than expected. It suffers from low quality, poor communications, a lack of management understanding, and a lack of structure. Unfortunately, these problems are often visible to the customer.

Unnecessary steps, errors, and changes make the workflow expensive. If the waste was removed, the process would become cheaper and quicker.

Unlike the manufacturing process, for which the workflow is defined in detail, the workflow in the engineering process is generally not well structured, except in those areas where it is bureaucratic. In these areas, it has far too much overhead structure. There is no real quality assurance or quality control in the engineering workflow—or so it would appear from the number of engineering changes that are due to errors, not enhancements. Often, there are no formal procedures for estimating the engineering time and cost. In the absence of any clear structure, it is difficult to optimize the workflow, with the result that estimates of, for example, lead times will invariably be longer than necessary. The lack of workflow structure makes it difficult to manage the time and cost of a given project, with the result that engineering projects are often late and over budget. In many cases, these problems are reduced by the skill and experience of the engineering managers involved. However, the opposite is generally true in cases where management does not have a good understanding of the workflow. More overhead is then added, in the form of unnecessary management reports, which are sent to people who neither understand their contents nor are capable of acting on them.

Some of the most serious communication problems occur at the borders between functional organizations. Design engineers pass manufacturing engineers designs that cannot be produced. The design has to go back for correction. Engineering changes costing thousands of dollars result. The engineering/marketing, engineering/finance, and engineering/field frontiers are also potential sources of problems. Sales people offer customized versions without knowing if it will be possible to make them profitably. Design engineers are unable to get cost information from the finance function. Design engineers do not receive field reports and then design existing problems into new products. Maintenance requirements are not taken into account during conceptual design. The traditional engineering workflow does not encourage users in different functions to communicate freely to prevent these problems.

Many top managers are out of touch with the engineering process. They may feel left out by the computerization of engineering. They may feel that engineering workflow is not a top management issue. They often see mergers and acquisitions, and not improvements to company operations, as the way to improve shareholder

Expensive

Slow

Late

Wasteful

Bureaucratic

Low Quality

Poor Communications

Unstructured

Poorly Controlled

Not Understood By Management

EXHIBIT 3.2   Some characteristics of the traditional engineering workflow

value. Before going ahead with EDM/EWM system implementation, a lot of work may have to be done to change these attitudes.

3.5 WORKFLOW IMPROVEMENT TECHNIQUES

To overcome some of the problems associated with the traditional engineering workflow, various techniques have been proposed. Among them are those that aim for a general improvement in quality and a reduction in waste. These include total quality management (TQM) and just-in-time (JIT). Other techniques, such as those called simultaneous engineering, concurrent engineering and focussed product team approach, aim to improve workflow through the use of multidisciplinary teams, particularly in early stages of the workflow.

Another set of techniques includes design for manufacture (DFM) and design for assembly (DFA). DFA aims to reduce the cost of assembly by simplifying the product and process through such means as reducing the number of parts, reducing or eliminating adjustments, simplifying assembly, and ensuring that products are easy to test. For example, tabs and notches in mating parts make assembly easier and also reduce the need for assembly and testing documentation. A reduction in the number of parts will help reduce inventory and inventory management activities. Modular products that make use of common parts allow the variety required by marketing, while limiting the workload on the manufacturing function. Companies using DFA techniques have reported reducing the number of parts by up to 85 percent, and assembly time and number of assembly operations by similar percentages.

DFM is oriented primarily to parts rather than products. DFM aims to eliminate unnecessary and expensive features on parts. It aims to eliminate those features that are difficult to manufacture. DFM helps prevent the unnecessarily smooth surface, the radius that is unnecessarily small, and the tolerances that are unnecessarily high.

DFA and DFM need to be carried out at the conceptual design stage before major decisions have been taken about product and process characteristics. One problem that they face is that, at such an early stage in the design, there is often insufficient detailed information available.

Other techniques, some of them overlapping, have been developed to improve designs and the workflow. Standardization of parts and processes can lead to use of group technology techniques, which exploit similarities in products and processes. Early manufacturing involvement brings manufacturing engineers into design activities that take place early in the workflow, rather than bringing them in after the designers have finalized a product that will be difficult or impossible to manufacture. Design engineers can be trained to have a better understanding of materials and processes. Personnel from different functions can be located together. Rule-based axioms of good design, bringing together properties of successful designs, can be followed. Empirical guidelines are available and can be followed. FMEA (failure mode and effects analysis) can be applied. Experiments based on the Taguchi methodology can be carried out (e.g., to identify which parameters are most critical to a particular product's manufacture). Quality function deployment can be introduced. Software walk-throughs can eliminate errors and misunderstandings.

The above overview shows the variety of techniques and approaches available to help improve the workflow. Since EDM and EWM are closely linked, any serious attempt to introduce EDM should raise questions about the workflow. One or more of the above techniques may then need to be considered.

3.6 JUST-IN-TIME

JIT is usually only thought of as an improvement technique for the shop floor, but it can be applied to all activities in the engineering workflow, including those that traditionally have taken place in the engineering department. JIT aims to eliminate non-value-added activities. The results of applying JIT on the shop floor include reductions in stocks, delays, overruns, errors, and breakdowns. These vices can be found, in one form or another, throughout the engineering workflow. JIT techniques can be applied to remove this waste from the engineering workflow before automation is introduced.

JIT focuses on continuous flow. On the shop floor, in a non-JIT environment, manufacturing is carried out in a set of discrete steps with WIP (work in process) being moved to and from stores at each step. This leads to all sorts of problems, such as long manufacturing cycles, damaged goods, unnecessary transportation of goods, and unnecessary storage areas. A parallel can be drawn with the traditional engineering flow, which is also unnecessarily split up into a set of discrete steps, with much paper and many unfinished projects in various states of progress at several stages of the process.

On the shop floor, there are many non–value-adding activities, such as unpacking, inspecting, returning, storing, material handling, and rework. Exhibit 5.4 (Chapter 5) shows more than 50 common actions on engineering data. It is surprising to see how many of these actions (such as accessing, approving, archiving, backing up, capturing, changing, collecting, controlling, communicating, and copying) do not add value to the data.

Engineering activities should be reviewed to see how much time is spent adding value and how much time is spent on non–value-adding activities. Surveys showing that drafters spend more time looking for information than using it imply that there is enormous waste in some engineering activities. Engineering change management is a clear candidate for review, as are manual reentry of data, transportation of paper drawings, and office layout.

Changing the workflow changes the information flow. On the shop floor, material flow is all-important. In engineering departments, the "material" is information. Information flow is all important. Extrapolating from the introduction of JIT on the shop floor, the application of JIT to an activity in the engineering workflow would start with an analysis of data management, data quality, data flow, data communication (both internal and external), and data preparation. The workflow load and the workflow initiation procedures would also need to be examined.

3.7 SIMULTANEOUS ENGINEERING

Simultaneous, or concurrent, engineering brings together multidisciplinary teams that work together from the start of a project to get things right as quickly and as early as possible in the engineering workflow, with the overall intention of getting a quality product to market as soon as possible.

Sometimes, only design engineers and manufacturing engineers are involved together in concurrent product and process development. In other cases, the cross-functional teams include representatives from purchasing, marketing, accounting, the field, and other functional groups. Sometimes, customers and suppliers are also included in the team. Multidisciplinary groups acting together early in the workflow can take informed and agreed decisions relating to product, process, cost, and quality issues. They can make trade-offs between design features, part manufacturability, assembly requirements, material needs, reliability issues, and cost and time constraints. Getting the design correct at the start will reduce downstream difficulties in the workflow. The need for expensive engineering changes later in the cycle will be reduced. The overall time taken to design and manufacture a new product can be substantially reduced if the two activities are carried out together rather than in series.

Simultaneous engineering is a form of engineering workflow management. Among its aims is parallel execution of many of the activities that were previously carried out in series. Examples from companies using simultaneous engineering techniques show significant increases in overall quality, 30- to 40-percent reductions in project times and costs (Figure 3.1), and 60- to 80-percent reductions in design changes after release. These results demonstrate that EWM, the management of the engineering workflow, is important. EWM is as important as EDM, the management of engineering data. To make concurrent product and process design a real success, all the necessary information concerning products, parts, and processes has to be available at the right time. A lot of partially-released information has to be exchanged under tightly controlled conditions. EDM provides this type of environment. EDM and EWM must be addressed together. There is a danger when implementing EDM that the very visible problems of data management, such as the existence of many islands of automation, will lead to a solution that only addresses engineering data and computer systems. This tendency must be avoided, with EDM and EWM being addressed on equal footings.

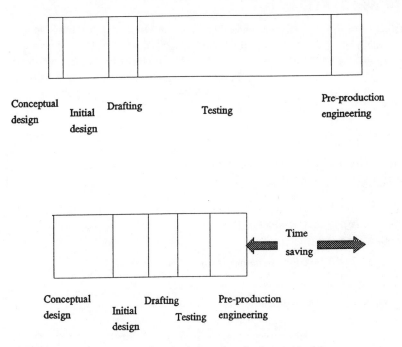

FIGURE 3.1   Simultaneous engineering helps reduce development lead time

4

Functions of an EDM/EWM System

4.1 EIGHT BASIC COMPONENTS OF AN EDM/EWM SYSTEM

From the issues and requirements described in the previous chapters, the functions required of an EDM/EWM system can be identified. There are eight basic components of an EDM/EWM system (Exhibit 4.1). The name of two of these components includes the word "Manager." In this context, "Manager" implies a software module with management functionality, not a person in a management role.

The first component is the Information Warehouse. Information is stored in the

---

Information Warehouse

Information Warehouse Manager

Infrastructure

System Administration Manager

Interface Module

Product And Workflow Structure Definition Module

Workflow Control Module

Configuration Management Module

EXHIBIT 4.1   Information Warehouse

Information Warehouse. A large volume of information needs to be stored. The information can be on any medium. It can be text, numeric, or graphic. If it is graphic, it can be in vector or raster form. The Information Warehouse need not be physically centralized.

The Information Warehouse is managed by the Information Warehouse Manager. This is the second component of the EDM/EWM system. The Warehouse Manager controls and manages the information in the Information Warehouse. It is responsible for such issues as data access, storage and recall, information security and integrity, concurrent use of data, and archival and recovery. It provides traceability of all actions taken on data.

The EDM/EWM system requires a basic infrastructure of a networked multi-vendor computer environment. (In this context, infrastructure means basic computer and communications hardware and software.) This may include other data management systems. The infrastructure includes a range of graphics terminals, printers, plotters, and other devices. The network will be used for both local and wide area communications, and for both short transfers (such as messages) and long transfers (such as files).

The fourth component of the system is the System Administration Manager. The complete system is under the control of the System Administration Manager. This is used to set up and maintain the configuration of the system, and to assign and modify access rights.

Users and programs access the system through the Interface Module. This provides a standard, but tailorable, interface for users. The Interface Module supports user queries, menu-driven and forms-driven input, and report generation (Exhibit 4.2). It provides interfaces for programs such as CAD, document scanning, electronic publishing, and MRP 2.

The structure of the information and processes to be managed by the EDM/EWM system is defined by the Product and Workflow Structure Definition

User details	Work structure
User privileges	Work item lists
Department details	Work item history
Data identifiers	Work item relationships
Data locations	Work status
Data structures	Messages
Project structure	Revision/effectivity dates
Project status	Authorization details
Standards	Exceptions
Procedures	Change details

EXHIBIT 4.2   Typical EDM/EWM system reports

Module. The workflow is made up of a set of activities to which information such as resources, events, and responsibilities can be associated. Procedures and standards can also be associated with activities.

Once initiated, workflow needs to be kept under control. This is the task of the Workflow Control Module. It controls and coordinates the engineering process. It manages the engineering change process and provides revision level control.

The exact structure of all products in the system is maintained by the Configuration Management Module.

4.2 THE INFORMATION WAREHOUSE

The Information Warehouse stores information regardless of its medium (e.g., paper, tape, disk) or physical location (e.g., in engineering, in manufacturing, on-site, off-site). It can handle all the information in the company, thus permitting centralized control of distributed data. Information only has to be entered once into the Information Warehouse. All information is indexed and traceable, and can be searched for. The Information Warehouse acts as a single source for all engineering information. This does not mean that all information must be physically in the same place.

The Information Warehouse stores all types of data. Information can be of varying sizes and formats. Information may be text, numeric, or graphic. It may have been created internally or externally. Information will be in various states (e.g., in-process, in-review, released) depending on its position in the process. Information will have various structures. The Information Warehouse can also store different alternatives and versions.

Apart from product definition data, the Information Warehouse will also store information such as relationships, workflow models, and product configurations. It will store computer programs and technical manuals. It will store procedures and standards. For a given assembly, for example, the Information Warehouse may store information on workflow models, parts in the assembly, relations between the parts, CAD models, drawings, process plans, tooling drawings, test results, and field information. Similarly, information on hierarchical structures, such as car model and corresponding power train and transmission, can be stored in the Information Warehouse.

It must be possible to implement the Information Warehouse without converting all existing data to new formats.

4.3 THE INFORMATION WAREHOUSE MANAGER

The role of the Information Warehouse Manager is to store incoming information securely and with integrity, to provide controlled access, and to protect product information. The Information Warehouse Manager manages the Information Warehouse. It controls and guards the information in the Information Warehouse. The Information Warehouse Manager allows data to be entered into the Information Warehouse and retrieved. Once retrieved, it can be transferred, modified, and copied.

The Information Warehouse is sometimes referred to as a repository or a library. The Information Warehouse Manager is sometimes referred to as a librarian module.

The Information Warehouse Manager works in a multi-vendor distributed computing environment and a multi-organization, multi-company environment. It must be able to control, for example, a text file on one vendor's computer in one company that will be required by a CAD user on another vendor's workstation in a supplier. The Information Warehouse Manager must be able to keep track of information outside the company, for example, with suppliers.

To support a variety of users and engineering tasks, the Information Warehouse Manager allows for multiple views of data and multiple levels of data.

The Information Warehouse Manager provides check-in/check-out facilities for individual files and sets of files. It is used to set up and maintain parameters describing data characteristics. It stores and makes available information under allowed access conditions. It provides access to information through a range of permission levels. These allow access to be controlled by a variety of criteria such as user, product, project, group, device, state of information, and type of information.

The Information Warehouse Manager limits access to authorized users. Data can be given a range of classifications from user-private to public. Some users will have view-only rights. Some will be able to copy, others to read and write. At various, well-defined times in a project, these rights may change. Once design information has been released to manufacturing, individual users will no longer be able to modify it. The access conditions may change as the data moves through the product life cycle. Depending on its status, data may be read-only. To maintain integrity, multiple simultaneous update will be prevented. The Information Warehouse Manager can provide an audit trail of all action taken on data.

The Information Warehouse Manager supports private data bases (single-user), project data bases (multi-user, linked to project lifetime), and product data bases (multi-user, existing parts). It can manage information stored in simple files or in hierarchical or relational databases. It can manage information on paper.

The Information Warehouse Manager provides security information on all unauthorized attempts to access data. It automatically backs up the information it receives and is able to recover all information lost as a result of computer or human problems. It has responsibility for systematic and on-demand archival. This may be to electronic or traditional media.

4.4 INFRASTRUCTURE

The basic infrastructure of the system includes computers and a communications network. The system runs in a multi-vendor computer environment. A variety of data-creation systems such as CAD/CAM, document scanning, electronic publishing, structural analysis, and process planning will run on these computers. The infrastructure will probably include engineering workstations and personal computers, as well as other graphics devices. These may range from intelligent 3-D graphics terminals in the engineering function to less sophisticated view-only terminals on the shop floor. There may be some data management systems, such as relational data base management systems, in the environment. The infrastructure also includes other input devices, such as scanners, and output devices, such as printers and plotters.

The communications network will include both local area networks (LANs) and wide area networks (WANs), so that information can be communicated both on site and between sites. Both short "messages" and long data "files" will have to be transmitted on the network. A message may be only a few words long. On the other hand, the volume of a data transfer, such as a product description file, may be much longer. It may even be several MB long.

The infrastructure must be able to handle electronic messages, some generated automatically as a result of events occurring, that need to be distributed to system users who may be based locally or may be far away. They may even be on supplier or customer sites. Electronic messages will inform users and managers that an event has occurred and that work on the following task should now proceed. A message could, for example, inform a supervisor that a design has been completed and should now be reviewed. The infrastructure must be able to handle the transfer of long data files. A company may want to electronically transfer product definition files of many MB to a supplier.

4.5 SYSTEM ADMINISTRATION MANAGER

The System Administration Manager is a module that allows the initial configuration and environment of the system to be described. This module will also be used to handle the changes that will occur in the environment.

The specifications of the initial configuration will cover the computers, data bases, networks, systems, workstations, plotters, printers, and other terminals within the environment for which the System Administration Manager will be responsible.

The System Administration Manager will be used to define users and applications in the system, and to define and modify the access rights of individual users.

4.6 INTERFACE MODULE

People and systems will want to communicate with the EDM/EWM system. Suitable interfaces will be provided for both classes of user.

In some cases, people will want to directly access the data management system from a terminal without going through another system. They may want to directly query data held by the system. They may want to look at a part attribute or a workflow description. Users may want to know what work they should do next. They may want to check on existing parts or look at test results. The user interface needs to support queries of many different types as efficiently as possible. The interface should be common to all the graphics devices in the configuration.

In other cases, the "users" of the system will be computer programs that store, create, modify, process, or otherwise make use of data managed by the system. An efficient and secure interface is needed for access by these systems. The systems could include CAD/CAM, document scanners, software development systems, technical publication systems, business systems, and other data management systems. They could also include wordprocessing and spreadsheet systems. They may exchange large or small volumes of data with the system. The data could be on the same computer as the program, or elsewhere.

The user interface should be easy to understand and use without lengthy training. People from many functions will make use of it. It should be suitable for casual users, yet also offer efficient facilities for frequent users. The interface should include an online help facility. The user interface should include both menu-driven and forms-driven approaches. Users should be able to tailor both menus and forms to their own requirements.

The interface should offer report generation facilities. Again, it should be possible to tailor these to user requirements. Some users may want to view or print out information on products and parts. Others may want to generate reports on project status, change history, or attempts to gain unauthorized access to the system.

4.7 PRODUCT AND WORKFLOW
STRUCTURE DEFINITION MODULE

The Product and Workflow Structure Definition Module is used to define the initial structure of a product and the corresponding workflow. It also allows the structures to be detailed as work progresses. The product structure defines the information requirements of a product throughout its life cycle. There will be various generic classes of information, such as assembly drawings, part drawings, NC programs, and user manuals. Each class of information will have characteristic attributes and tasks. The product structure describes the information that is needed or is produced at each phase of the workflow. The workflow is defined as a set of tasks, characterized by resources, events, associated information, responsibilities, decision criteria, procedures to be used, and standards to be applied.

The product structure and the workflow structure are closely linked. For each group of products, there will be a specific product structure and a corresponding workflow structure. The product structure defines all the information describing a product or part. Some of this information will be created or used at each step of the workflow. New product structures can make use of parts of existing structures. In some cases it will be easiest to start at the beginning of a process and define the information that is to be created at each step of the process. In other cases it may be easier to start at the end of the process (e.g., with a bill of materials) and work backwards, identifying at each step the information that would be necessary to generate each information item. Initially, this task is very difficult, and a variable degree of detail is essential for the structure so that work is not held up by the unavailability of detailed information on the structure. The product structure has to be sufficiently flexible so as to be able to handle changes to information structures and items. The Product and Workflow Structure Definition Module should offer the possibility to add or associate information when enough information is not available to define a product, workflow, or relationship.

The workflow may be described for the entire product life cycle or for individual processes. The level of detail needed to describe the workflow is variable. The workflow may be described either by starting with the end product and working backwards to the beginning, or by starting with a blank sheet and working forward to the end product. The workflow can be split up into projects. In turn, these can be broken down into phases and steps. A workflow structure may be built up from existing steps, or a new structure may be built. In the same way that the product structure has to be sufficiently flexible to handle changes to the product structure, the workflow structure has to be able to handle changes to workflow.

The workflow structure will be used to control the engineering workflow. It has to include all of the information required to make this possible. It will become the source of work statements, defining the activities to be carried out and the resources to be used. For example, it should specify which users should be involved in the design of a particular part, the systems they should use, the information they will need, the information they should produce, the procedures they should follow, and the approval process. The approval process work statements should specify the roles and authorities of the individuals involved, the rules and requirements for sign-off, and the process to be followed if sign-off is refused.

During a process, "events" occur. An event marks the end of one activity and the beginning of another activity. Events need to be identified and included in the workflow definition. Completion of a design and initiation of a periodic maintenance activity are typical events. The activities that lead to and follow each event should be specified. The messages that should be communicated when the event occurs, or if does not occur within a specified time, have to be defined in the workflow definition.

The review, approval, and release process has to be defined. The engineering change process has to be modelled. The various steps, reviewers, approvers, and sign-off procedures have to be defined. The definition should include hierarchies so that, if an activity does not occur, it will automatically be passed to the next highest authority. Since many parts of the process are similar or repetitive, some workflow structure elements will be repeated throughout the overall workflow. The creation of change requests and orders is a typical example. Automatic process sequences can be set up to handle tasks such as the provision of copies to individuals named on a distribution list.

Once the product and workflow structures have been defined, the EDM/EWM system can manage data and workflow. Formal description of the workflow will give management an opportunity to identify unnecessary activities, as well as those that could be run in parallel. Eventually, knowledge-based systems will be able to take over this activity.

4.8 WORKFLOW CONTROL MODULE

For a particular module of work, the product structure and the workflow structure will have been defined by the Product and Workflow Definition Structure Module. The Workflow Control Module manages the workflow of the various activities in progress, and monitors progress. It controls the progress of projects in an event-driven mode. It maintains status information on ongoing projects.

Once initiated for a particular workflow/product pair, the Workflow Control Module has complete control of the process. It assigns tasks to individuals, informs them of the resources to be used and the procedures to be followed, initiates the associated actions, and maintains status information. If necessary, the Workflow Control Module can remind users of standard operating procedures, and can check that standards information is accessed. It distributes data and documents to the individuals as needed. When the task is finished it can request a review or promote the design and initiate the next step of the process. It can enforce promotion rules. If the person responsible for the next step is absent, it can automatically pass the work to the most suitable replacement or the next highest authority. It manages the review, approval, communication, and archival of information.

Automatic process sequences handle such tasks as providing copies to individuals on a distribution list. Following the rules specified during the definition of the workflow and product definition structures, the Workflow Control Module controls versions and manages the engineering change process.

The Workflow Control Module, on the basis of the workflow and product structures, ensures that all necessary information is available before releasing parts to manufacturing.

The Workflow Control Module monitors the occurrence of events. When an event occurs, it initiates a previously defined set of activities. If an event does not occur at the expected time, the appropriate level of management will be alerted.

When an event occurs, the Workflow Control Module sends, in accordance with the defined structures, the appropriate messages. Engineering change requests and change orders can be rapidly transmitted to inform all interested parties of impending and actual changes. Interested parties can be informed of upcoming events. The Workflow Control Module can notify other people that a change has been requested. It can initiate messages based on parameters captured at each step. It can notify downstream users that modifications have been made to upstream information.

The Workflow Control Module will keep status information up to date, and ensure that design information is handled as planned. At any time, the Workflow Control Module will be able to display the exact status of each process it is managing. It can track and report the status of tasks in process. It can produce progress reports at specified times, showing, for example, how much leadtime has been used up. The Workflow Control Module maintains an audit trail of activities relating to the process.

Performance analysis can be carried out on engineering activities and on information access. The impact of proposed changes can be analyzed and assessed. Resource-level loading can be coordinated and schedule visibility maintained for all related tasks.

Once set up for all projects in the organization, the Workflow Control Module should be able to promote parallel, rather than serial, workflow, thus reducing lead times. Increasingly, the Workflow Control Module should take on an "expert" role, checking that input is legal and consistent and that correct procedures are being followed, and automatically organizing activities so as to minimise lead time.

4.9 CONFIGURATION MANAGEMENT MODULE

The configuration items include all the technical documentation necessary to specify, build, test, install, operate, and maintain the product. This will include information in the form of specifications, drawings, lists, programs, reports, and manuals. The Configuration Management Module is used to describe the exact configuration of a particular product throughout its life. It relates components, subassemblies, and assemblies. It supports multiple assembly levels, multiple hierarchies, and multiple membership. The Configuration Management Module maintains a complete history of the product through design, manufacture, and delivery to field use. The status of all information (e.g., in-process, in-review, released) is maintained by the Configuration Management Module. It maintains the configuration of a given end product by managing all the information needed to produce the part. For a given product, for example, it may maintain information such as bill of materials, goes-into lists, assemblies, relations between the assemblies, CAD models, drawings, analysis results, parts lists, process plans, NC programs, tooling drawings, test results, and field information. The Configuration Management Module can distinguish between the as-designed, as-planned, as-built, as-installed, and as-maintained configurations of the product.

The Configuration Management Module maintains exact configuration information on each individual product. It supports multiple versions and alternatives of data. It takes account of engineering changes. It maintains information about the relationships between information, such as the creation of one file from another.

The Configuration Management Module offers the possibility to navigate product structure by paging down and traversing the workflow and information structures. It allows information to be accessed in many ways, such as by model number and by part number.

5

The Need for EDM/EWM Systems

5.1 TEN ISSUES RESOLVED WITH EDM/EWM SYSTEMS

At first sight, EDM/EWM systems may just appear to be the answer to the fairly well-defined problem of how to manage large numbers of CAD/CAM files. However, further examination shows that this is not so. EDM/EWM systems respond to a large number of complex and currently unresolved issues that are all related to the problems of managing engineering information and engineering processes. Some of these issues appear as business problems and some at the level of the engineering function. Some are management problems and some are very closely related to engineering data itself. The problems can be grouped, from the technical viewpoint, into ten major categories (Exhibit 5.1).

5.2 DIVERSITY AND VOLUME OF DATA

The sheer volume of data and the diversity of its physical support media (Exhibit 5.2) make engineering data difficult to manage. There is clearly sufficient difference between data on traditional media, such as paper, and data on electronic media, such as an optical disk, for them to require different management techniques. However, there are also great differences between the various traditional media, such as paper and aperture cards. Similarly, due to the diversity of computers, operating systems, storage devices, and storage techniques, there are many differences between the various electronic media.

Computers have been used in the engineering environment for nearly 50 years.

Diversity and Volume of Data

Multi-User, Multi-Organization Environment

Multi-Application, File-Based Environment

Multiple Data Definitions

Multiple Representations of Data

Multiple Versions of Data

Multiple Relationships

Meaning of Data

Life Cycle Data

A Complex but Logical Environment

EXHIBIT 5.1   Ten issues resolved with EDM/EWM

Numerical control of machines has been applied for over 40 years. The earliest attempts at what is now known as CAD/CAM took place over 30 years ago. It could seem surprising that it is only now, in the 1990s, that the need for engineering data management is really becoming apparent. However, the earliest applications of computers in the engineering environment were point solutions, creating "islands of automation." Self-management was possible on small, isolated islands. While an island remained small enough, the amount of data it required and generated could be managed manually. It was only towards the end of the 1980s, as the number of islands grew and individual islands became much larger, that the data management problem really became apparent.

Since the creation of files ranging from 50 KB up to 1 MB or even 10 MB only requires a few minutes, it does not take long, even in small companies with only a few users of computer-based engineering data, for manual data management techniques to break down. Users will not be able to find data. Data will not be secure. In larger companies, with several hundred engineers, many gigabytes (GB) of data may be created and accessed each week, making the problem far worse.

The volume of data in the engineering environment is of itself a problem. Estimates for medium-to-large companies foresee data volumes exceeding 1 million GB. There is currently no single electronic storage device that can handle such a volume of information. Of necessity, therefore, data will be stored in several locations. It will be difficult for a user to know where to store data. In what format

Paper	Microfilm
Paper Tape	Core Memory
Vellum	Magnetic Tape
Mylar	Magnetic Disks
Aperture Cards	Optical Disks
Microfiche	

EXHIBIT 5.2   Storage devices for engineering data

should it be stored? What level of security will be appropriate? How will a user be able to find out where data was stored in the past? How will a user be able to find out which format was used to store it?

A company's engineering data represents its collective know-how. As such, it is a major asset and should be used as profitably as possible. Too many companies ignore their engineering data. If $100,000 goes astray in their financial systems, there is a major panic. If $10,000,000 goes astray in their engineering systems, there is generally no panic at all, since most people in the company are completely unaware of the loss. Many managers find it difficult to put a value on their engineering data. Top management, in particular, is rarely aware of the extent to which valuable data is ignored and misused.

The scope of engineering data is wide (Exhibit 5.3). There are various types of data. There is text data (specifications, schedules, process plans, manuals, project plans), numeric data (descriptive geometry, formulae, results of analytic experiments and calculations, computer programs), graphics data (photographs, drawings, sketches), and voice data. Within each of these types of data, there may be differences. Among the computer programs that need to be managed are both those that are linked to products (programs developed to be used within the company's products) and those that are linked to processes (programs, such as MRP 2 and CAD/CAM, that support the company's operations). Some of the graphics will be in vector form, some will be in raster form.

Data is processed in many ways (Exhibit 5.4). The best way for processing and managing any one of the different types of data will not be the best way of processing and managing another. If users want optimum performance, they will suffer from not having a common approach. If they prefer a common approach, they will not have optimum performance.

Some of the data will be on paper. The paper will be of various sizes, probably ranging from the U.S. A-size (European equivalent A4) to the U.S. E-size (European equivalent A0). Some data will be on mylar or microfilm. Some will be on

Analysis results	Parts lists
Assembly drawings	Plant layouts
Bills of materials	Process plans
CAD and CAM geometry	Process recipes
Circuit layouts	Product designs
Classification numbers	Product specifications
Computer software	Project plans
Cost breakdowns	Quality data
Customer requirements	Repair instructions
Detail drawings	Results of experiments
Design specifications	Sales information
Engineering change orders	Schematics
Engineering change requests	Set points
Facility drawings	Simulation models
Feature data	Sketches
Field data	Spares catalogues
Field service manuals	Specifications
Fixture drawings	Standards
Flow diagrams	Technical documentation
Flowcharts	Technical features
Formulae	Technical manuals
Installation drawings	Test results
Item numbers	Test vectors
Listings	Timing charts
Maintenance documentation	Tolerances
Machine library	Tool lists
Material descriptions	Training material
Models	User manuals
Notes	Wire lists
Numerical control programs	Word processed text files
Numerical control tool paths	Welding drawings

EXHIBIT 5.3 Examples of engineering data

paper tape and some on magnetic tape. Some will be on magnetic disk, some will be on optical disk. Tapes will range in size from small cassettes up to many MB magnetic tape reels. Disks will range in size from a few MB up to several GB.

Many different computer programs (Exhibit 5.5) are used in the engineering environment. They create and work with engineering data in different ways. In addition to the previously mentioned CAD/CAM and NC systems, there are finite element analysis, aerodynamic analysis, process planning, technical publishing, word processing, test, and many other types of system. Each one of these will probably be an island of automation, with its own specific approach to data management. Each will primarily address the function it is supposed to perform, such as geometry definition, with data management being a secondary (and ineffectively implemented) function.

To process data, the engineering function will be using computers of different sizes (Exhibit 5.6) and from different vendors. There may be a supercomputer and one or more mainframes. There will be many minicomputers and microcomputers. There will be a variety of personal computers and engineering workstations. There will be some graphics terminals without significant computing power. Some of the computers will be linked together on one network. Others will be linked together on other networks. Some will be stand-alone. The hardware alone may come from 5, 10, 20, or even 50 vendors.

The computers will run under a variety of operating systems. Some of these will be proprietary (i.e., not standardized). Others will, in principle, conform to a standard, such as UNIX. However, even those that are, in principle, standardized may have minor differences, particularly between different versions and releases.

5.3 A MULTI-USER, MULTI-ORGANIZATION ENVIRONMENT

Engineering data will be used by many people (Exhibit 5.7). These people may be in different functions and locations of a company. They may be outside the company. They may be working for a supplier or a partner, or they may even be the final customer of the company's product. Data has to be made available to all these people. At the same time, data must be protected against unauthorized access.

There will be many users of data. A given piece of information may be created by a design engineer, analyzed by another engineer, drafted by someone else, checked by a supervisor, and scrutinized by a manufacturing engineer even before it is accepted as potentially useful. These people may be in the same building or in the same plant, but they could also be in locations in different countries or even on different continents.

Data will be spread between different organizations. Clearly, a lot of information will be created in the engineering organization, but information will also be created and used in the manufacturing, marketing, finance, sales, and maintenance organizations. Some of the data will probably be with the design engineer. Some will be with the manufacturing engineer. Some will probably be with the EDP department that looks after the MRP 2 program. Some data will be on the shop floor. Some will be with the customer.

There will probably be a lot of redundant data. The design engineer will have a copy of some of the information that is used in production planning. Copies of

Accessed	Manipulated
Added	Modified
Altered	Moved
Approved	Ordered
Archived	Output
Backed Up	Prepared
Burnt In	Printed
Captured	Processed
Changed	Protected
Checked	Punched In
Collected	Punched Out
Controlled	Read
Communicated	Recorded
Copied	Released
Created	Removed
Defined	Renamed
Deleted	Restored
Destroyed	Retrieved
Digitized	Reused
Displayed	Reviewed
Disseminated	Saved
Distributed	Secured
Drawn In	Selected
Drawn Out	Sent
Duplicated	Shared
Edited	Sorted
Entered	Spoken In
Exchanged	Stored
Filed	Traced
Found	Tracked
Grouped	Transferred
Handled	Transmitted
Indexed	Typed
Input	Unloaded
Inserted	Updated
Loaded	Used
Located	Viewed
Lost	Written In
Managed	Written Out

EXHIBIT 5.4   Actions on engineering data

Aerodynamic analysis	Kinematic analysis
Assembly	Machine allocation
Bill of materials processing	Machine simulation
Circuit layout	Manufacturability analysis
Circuit simulation	Maintenance systems
Computer-aided design	Numerical control
CAD/CAM	Numerical control programming
Computer-aided drafting	Order processing
Computer-aided engineering	Production planning
Computer-aided packaging	Project management
Computer-aided publishing	Process planning
Computer-aided software engineering	Purchasing
Configuration management	Quality assurance/control
Costing	Quotations
Crash analysis	Risk analysis
Data conversion	Reliability analysis
Data exchange	Sales systems
Document scanning	Shipping
Electronic mail	Shop floor systems
Engineering calculations	Simulation
Engineering change control	Spreadsheets
Forecasting	Structural analysis
Graphics	Technical publishing
Inspection	Test systems
Inventory control	Word processing

EXHIBIT 5.5   Systems that create or use engineering data

some of the information used to generate NC programs will be kept by both design engineers and manufacturing engineers. Maintenance engineers may want to keep "their" drawings close at hand so that they can respond quickly to urgent customer calls. Nearly all of this information will be a copy of information stored elsewhere in the company.

Traditionally, it has been made clear, through organization charts, which human resources belonged to which part of a company's organization. It has been less clear which information belonged to which part of a company's organization. Even within a particular part of the organization, such as engineering, it has not been clear who are really the owners of information. Owners have both rights and responsibilities. Designers, analysts, drafters, supervisors, and managers will all have their own ideas as to ownership of information. They will probably be willing to defend their "property" if anyone suggests that the information should actually

Communication controllers	Minicomputers
Engineering workstations	Personal computers
Graphic devices	Programmable logic controllers
Machine controllers	Servers
Mainframes	Supercomputers
Microcomputers	Superminicomputers
Microprocessors	X terminals

EXHIBIT 5.6   Computers used to process engineering data

be owned by somebody else. They may be less willing to accept responsibilities, such as maintaining it properly, and making it available when needed by others. Outside the company, a similar ownership issue will arise as data is shared with partners, suppliers, and customers.

Users will be working on a variety of tasks. Depending on what they are doing, and their level of computer literacy, they will have different data usage and data management needs. Some will create data, and some will modify it. Others will only want to reference existing data. Many of the users will have an engineering background, but those outside engineering will probably come from other backgrounds, such as accountancy, marketing, and sales. Some of the users will be working with advanced concepts, such as the creation of data that will help take man out of the solar system. Others will need data to help them solve more down-to-earth problems, like finishing a part before the shift ends.

Different users will want to see different views of the data. They will want to see the view that is of interest to them. A manager may want to see current progress on all parts of a new project, but not details of the design itself. A project engineer may want to check an assembly of parts, but have no interest in stress analysis results. A drafter may only be interested in an individual part. A company may only want to give a partner a very restricted view of its overall database. The company and its partner may be using different computers and application programs to access the data. While the view of data that users may want to see is different, and the systems they use may be different, the underlying data must be the same.

While it is true that different users will have different requirements of a data management system, it is clear that they will also have some common requirements. For example, they may all want to make use of the same basic spreadsheet, text processing, and electronic mail systems.

At the level of the data itself, there is a need to provide better security, control, access, and protection. Although data is so valuable, it is often very easy for users to mistakenly destroy or lose sight of data. The need to provide full protection of data has to be addressed alongside the need to make data available. At different times in the product life cycle, data should have different levels of

Within the company

R&D/ENGINEERING

Designers	Process engineers
Mechanical engineers	Quality engineers
Electronic engineers	Software engineers
Drafters	Engineering managers
Manufacturing engineers	Structural analysts
Scientists	Laboratory technicians

MANUFACTURING

Manufacturing planners	Production schedulers
Machine operators	Manufacturing managers
Maintenance staff	Expediters

FINANCE/COMMERCIAL

Buyers	Accountants

MARKETING AND SALES

FIELD SERVICE/MAINTENANCE

Outside the company

SUPPLIERS	PARTNERS
CUSTOMERS	AUDITORS
REGULATORY AUTHORITIES	CO-MARKETERS
SUBCONTRACTORS	DISTRIBUTORS

EXHIBIT 5.7   Potential users of EDM/EWM systems

protection. Until a design is released, a design engineer may be free to modify it. After release, manufacturing engineers should not be allowed to make "improvements" without following the correct procedures. At times, the situation will be complicated by the need to share data among several users, each of whom has different access rights.

It is difficult for users to access data that is not readily available, and, rather than searching and waiting for it, they may prefer to ignore it or re-create it. For example, designers rarely have easy access to product cost data in MRP 2 systems and product quality data in field support systems.

5.4 A MULTI-APPLICATION, FILE-BASED ENVIRONMENT

Most of the application programs currently used in engineering activities are file based. They store data directly in files under the control of the computer's operating system. Another program on a different computer cannot easily access the data in these files. There are two reasons for this. The first is due to the operating system and the communications network, and relates to the difficulty of transferring data from a file under one operating system on one computer to a program on another computer running under another operating system. The second is that information about an object such as a part or product is not independent of the program that created it. The knowledge about the structure and meaning of the data in the file is often only in the application program that wrote the file (or in the head of the person who developed the program) and is not available to the other program. As a result, even if the latter program were able to access the data physically, it would not be able to understand it.

To overcome the problems of transferring data electronically between programs, it is often transferred manually. Companies often plot a drawing with one system, and digitize it manually into another system. In one aerospace company, it was found that data was being manually transferred like this through a chain of seven functions, including assembly drawing, analysis, detail drawing, tool design, NC programming, shop floor instruction preparation, and service manual development. Manual transfer of data introduces errors and wastes time. It can take so long that users may decide it is not worthwhile. Instead, they may work with out-of-date or incomplete data. For example, an engineer may not be able to get data directly from the costing system, and will develop a design without taking sufficient account of cost information. Sales staff may not respond to customer queries because it takes them too long to get to information.

Sometimes it is possible to develop software that extracts the required information from one program's file and reformats it into a structure that is acceptable to the second program. Initial development of this software, of course, takes time and money. In addition, each time that either of the programs is changed, more effort has to be wasted in maintaining the conversion software.

As many program developers are unwilling to inform program users as to the structure and meaning of the data in their files, data in a file often remains understandable only to the program that wrote it. Files "belong" to the programs that wrote them. For each island of automation, an "island of data" appears. Medium-sized companies may have 20 or more islands of data.

The users of application programs are of course much more interested in real objects, such as products and parts, than in the structure and format of data in files. Too often, though, the information on the products is only available after wading through and understanding many sets of files. Again, this represents a waste of time. "Bridges" need to be built so that information can be moved from one island to another (e.g., part specifications and engineering changes must be transferred from the CAD island to the MRP 2 island). Bridges often need to be company-specific and are time-consuming to build and maintain. Even when the bridges are in place, it will often be found that the information needs additional conversion, interpretation, and synchronization.

Each new application program has, unfortunately, the potential to become a new "island of automation" and to create a new "island of data." A company can overcome the problem, in part, by using only an "integrated set of applications" from a single vendor. In this case, some of the physical data transfer problems will be reduced, and the vendor may provide some means for improving the flow of information between the individual programs. In all but the smallest companies, however, it will not be feasible to buy only an "integrated set of applications" from a single vendor. Since most companies will have needs that cannot be met effectively by such a set of applications, other systems will have to be acquired to handle these needs. As a result, new "islands of data" will arise. In the future, as current technologies evolve, and new technologies are introduced, new "islands of automation" will appear. They may be of great importance to individual companies, who will acquire them even if they are not integrated. For the foreseeable future, companies will have to cope with incomplete integration, application-related files, and the resulting problems of working with information that is connected in real life but unconnected in the computer systems. These problems, through redundant data, redundant data entry, redundant conversion of data, and redundant programs lead to increased operating costs.

Even if the engineering function could overcome all these problems and consolidate all its computing and communications activities into one "island of automation," it would still face the problems of working with the "island of automation" in the manufacturing function, and with the engineering function in partner companies. It is unlikely that these companies would have chosen exactly the same "island of automation" solution. They could have chosen a different solution, or they might have decided to work in an environment that is not integrated.

As each application stores the information it requires in its own files, there will be overlap between the information stored by different programs. As an example, many applications will make use of product names and part numbers. Should these change, the corresponding application programs will have to be changed. This takes time and effort, and may introduce errors.

As well as storing data in their own specific ways, application programs have their own specific user interfaces. Their approach to common functions (e.g., math functions) is also specific to each application. Each time someone uses one of these programs, they waste effort in first learning and then remembering the specifics of the program. Standardization of user interfaces and basic functions is part of the overall approach to the management of engineering data and workflow. File-based application programs are, in part, a reflection of current computer technology, but they are also a reflection of the organizational environment in which they are used. In the past, organizations have tended to be departmental, and application programs matched the functionality and data needs of a particular department. As organizations change and "departmental walls are broken down," functionality and data needs change. Unfortunately, most application programs cannot easily change to meet them.

5.5 MULTIPLE DATA DEFINITION

When there is no standard definition of the data associated with a particular part or product, each user (and application program) can have a different definition of the data, and all the definitions can be different. Different functions may even use different part numbering systems. This leads to errors and wasted time and money, yet many companies have several different definitions of some data items. A CAD program may have one definition of a part. A part programmer may redefine it. A stress analysis program may use a third definition. In the bill of materials, the part may have another definition. It may be redefined for inspection, and again in assembly instructions. Unless the company has introduced strict procedures ensuring that all these definitions are equivalent, there will probably be minor differences between them. These will lead to confusion when modifications are made to the part or if an attempt is made to reuse the part in another product or design. Since the definitions are not identical, the result of a modification to one definition may not be the same as the result of the same modification to another definition. Special software may have to be developed and maintained to allow users to continue working with their own definition.

Often, two users will have conflicting definitions of the same object. As neither wants to appear to have the "wrong" definition, the data is "manipulated" so that both definitions can be maintained. This becomes a problem when data has to be transferred from one user or application to another. In some cases, part of the definition will be lost.

In other cases, something extra has to be added. In all cases, there is the possibility for errors and approximations to be made. This can only decrease the overall quality of the final product. The lack of a standard data definition causes further problems when changes are made by one of the users. The change may mean that the manipulation that was applied previously is no longer suitable. This can set off a chain reaction in other parts of the product life cycle, leading to more mistakes and wasted resources.

Similar problems arise with the use and management of data in libraries of standard parts. It has to be decided which data is to be stored in libraries, when it is to be created, who can access it, and, perhaps most importantly, who can modify it and when. Problems arise with references made in old products to standard data that is to be modified. In some cases it may be best to stay with the old standard. In others, it may be best to bring them into line with the new standard.

The level of data definition changes throughout a product's life cycle. In the early stages of the product's life, there is little data available. The level of detail increases as the product is developed and used. Once the product has been shipped to the customer, the level of detail required often falls. The definition of the product does not have to be identical at all stages, but it does have to be consistent.

Development of a standard definition of a data item needs to take account of the needs of the various users and application programs that make use of it throughout its life cycle. The choice and implementation of a standard data definition is time-consuming and complex.

5.6 MULTIPLE REPRESENTATIONS OF DATA

Different users will want to look at and work with different representations of data. Sometimes, the representations will make use of the same data type. At other times, even the data type will be different. In all cases, the representations must relate to the same underlying data. Modifications have to be made to the underlying data and not to the more superficial representation.

At various times, different users will want to work with different structures of data. Some users will want to work with different hierarchical levels within a given structure. Some will want to include the same part in different functional or hierarchical structures. Engineering, accounting, production management, and assembly may all have different requirements for bill of materials structures.

The structures and levels have to be consistent. At the lowest levels, the data will be in the form of bits and bytes. These may represent numbers, characters, sounds, and lines. In turn, these may represent geometric information or information about a material or a color. At a higher level, this information may actually represent a part, which in turn may be a component of an assembly such as a wing flap, which in turn belongs to a wing, which in turn belongs to an aircraft. Each level of the structure is of interest to particular users. The machinist drilling the holes for the bolts that attach the wings of an aircraft to the fuselage wants to know the exact position of the holes and any deviations during drilling, and is not interested in the aerodynamic qualities of the wing.

One user may need a bottom-up structure of the product, starting with nuts and bolts and small parts and then working through larger components and assemblies. Another user may require a top-down structure, starting with the complete product and then working down through the major assemblies.

Some users will be happy to work with two-dimensional data. An engineer laying out a single printed circuit board may not need to take account of the third (thickness) dimension. On the other hand, a stylist defining the shape of a car wing will require a full three-dimensional representation of data. Other users may need to work with both two-dimensional and three-dimensional representations, and want a modification to one of these to be reflected in the other.

A related issue is that of a drawing of a part on paper that is electronically scanned and then converted from raster format to CAD/CAM vector format. The three representations of the part are different representations of the same object. Each one can play a useful role in the product life cycle. An analyst may need to use the CAD/CAM model. A machinist may need a rasterized picture on a shop floor terminal. A maintenance engineer may have to make a major repair on a site, and may need to take a paper drawing to the site. Procedures have to be in place so that the different representations can coexist and so that any necessary modifications can take place. Any modification made to one of the representations has to be reflected in the others.

Under their own data management functionality, different application systems may store the same data in different formats. For example, one system may represent a circle by its center and radius, whereas another system may represent it by three points on its circumference. A third system might represent it by its center and two points on its circumference. Even systems that use the same representation may physically store the data differently. One system may store it in the order x-coordinate of center, y-coordinate of center, radius, whereas another system may store it in the order radius, x-coordinate of center, y-coordinate of center.

5.7 MULTIPLE VERSIONS

The engineering environment is typified by many versions and alternatives. Products are made with different models, versions, options, releases, and alternatives. Throughout the engineering process, designs change, components are modified, products are restructured, and project status is updated accordingly. Management tools, such as project management, configuration management, and engineering change control, are applied to maintain control.

The tendency to produce customized products in small batches increases the load on these systems. Whereas 30 years ago a car might have been produced in just a few variants, automotive manufacturers must now handle millions of variants. As product lifetime and time to obsolescence decrease, new models have to be brought out more frequently. The range of products increases and, as products are customized, the number of possible combinations of parts rises dramatically.

For the manufacturer, the environment becomes increasingly complex and hard to manage. As the number of potential product configurations increases, it becomes harder to keep track of the real configurations of individual products. It becomes harder to make sure that configuration documentation corresponds exactly to individual products. The customer requires the same after-sales service on a product that is effectively a one-off, as if the product had been produced as one of an identical batch of several thousand.

The systems that should support versions and changes have not evolved fast enough to handle increased customization and the rapidly increasing number of changes, variants, and versions. The engineering function in many companies is still trying to manage them with manual systems or with automated systems that are not integrated with the engineering systems, such as the CAD system, on which the changes occur. Manual systems are increasingly failing to control engineering drawings and to manage engineering changes and product configurations. Typical results are that products are late and after-sales service is poor and expensive since no one knows what components actually went into a customer's product. Customer-requested changes may be a real problem if they are so expensive that their real cost cannot be charged to the customer.

Those who create and make use of engineering information are faced with an ever-growing number of versions and alternatives of the information associated with their products. Users are often unsure if they are using the right version of the data. As the rate of change required by market forces increases, the information management systems they use come under pressure with the result that time is lost and mistakes are made. It becomes increasingly difficult to make sure that the information in these systems conforms to the reality of the engineering process.

The different versions of systems, such as CAD/CAM systems, are another potential source of problems. The data management capabilities of successive versions of a CAD/CAM system can be incompatible. The system vendor may well upgrade the system with the intention of providing better functionality and richer information content, but by doing so may create the situation where an earlier version cannot make use of all data created under the new version, and the new version can be limited in its ability to use data created under the earlier version. It may be necessary to "renovate" existing data when a new version is implemented.

A corresponding problem relates to the computer hardware and operating system at the heart of the CAD/CAM system. If past trends continue, hardware currently in use will not be in use in 20 years time, yet some companies will need, in 20 years, to access data currently being created. It could be difficult for users in the 21st century to re-create the present environment exactly, unless the company intends to archive its computers, operating systems, and programs, as well as its data.

When agreement has been reached as to the ownership and location of data, the question will arise of how to manage necessary copies of the data. An engineer working on a new product may want to build up a temporary, individual data base, containing copies of existing products that are normally stored in a central location under central management. Another engineer, having to make a major change to the design of one of these products, needs to be aware that the design is in use and to be able to signal other users that the previous version is unsatisfactory, that changes are necessary, and that these will not be available for several days.

A design engineer may be working on a new design. To help reduce lead times, a copy of the design may be sent to a manufacturing engineer before it is officially released. Time can be saved if work on NC programming can start before the design is finalized. However, procedures need to be in place to ensure that further changes to the design (which was not officially released) are transmitted to the part programmer.

Mistakes are expensive. The longer it takes to discover that the wrong part has been released to manufacturing, the more costs will be incurred and time wasted, as incorrect documentation is developed, the wrong tooling is developed, and wrong orders are placed with suppliers.

In the past, the lead time for many products was dominated by their mechanical component. Recently, the percentage of electronic and software components in many such products has increased rapidly. Revision time for these components, in particular for computer programs, can be very short. Unless versions are closely controlled, configuration management becomes a nightmare.

5.8 MULTIPLE RELATIONSHIPS

As well as working with the data, engineers and other users of engineering data are interested in the relationships between data. As well as managing data, it is also necessary to manage the relationships between data.

There are many types of relationships. There are relationships, for example, between products and parts, between parts and data, between one part and another, and between parts and workflow. There are relationships between the parts of one product and the parts of another product. There may be relationships between apparently separate projects. There are hierarchical relationships linking parts to a product. Bills of materials, parts lists, assembly drawings, and where-used lists contain information on such relationships. The various types of data (such as specifications, drawings, models, test results) supporting a product need to be related. A component may only fit a particular product, an engineering drawing correspond to a particular part, or an NC program correspond to a particular version of a part. There are relationships between activities. The design of a particular part may only be started when the design of other, related parts has been completed. Data needs to be linked both to its source and to derived data. It also needs to be linked to the processes that create and use it. Users need to be able to navigate through the various relationships and links.

In addition to the wide variety of relationships in existence, added complexity arises due to changes that occur in relationships as the engineering process takes place. For example, a part that was previously used in three products may, in the future, be used in a fourth product. A part that was originally expected to be made of metal may instead be made of plastic. Similarly, the relationship between logic changes and printed circuit board (PCB) designs needs to be maintained so that changes in the logic are reflected in the board. Parts of a design may be due not to structural or aesthetic reasons, but to the manufacturing process that will be used. A part that has been designed to be cast may have features that are completely unnecessary if in the future the part is to be machined.

Information needs to be maintained on the reasons why choices were made between competing designs. If the selection criteria can be retained, they can be used to help make better choices in the future. If the context within which a particular design was chosen is clear, it may be obvious how the part should be made.

Information concerning the completeness of a design needs to be maintained from the viewpoints of functionality, constituent parts, and necessary activities. If a particular part is replaced, it should be possible to identify what is needed to maintain completeness. A "missing" relationship may be as important as a "present" relationship. Comparative relationships, such as "duplicate," are needed, for example, to prevent development of identical products.

In the file-based environment, it is often difficult to maintain information on the almost limitless relationships between information in different files. For example, one file may contain analysis results on a part in a second file that was subjected to forces specified in a third file. The second file may also contain the geometry model of another part. A dimensioned drawing of this part may be stored in a fourth file. When the second part is modified, a fifth file may need to be created with the corresponding dimensioned drawing, but, because the first part may not have been modified, it may not be necessary to repeat the stress calculation. Probably, the file will contain no indication of whether or not the first part was modified, so the stress calculation will be repeated, generating a sixth file that is identical to the first file. In the real environment, in which there will be thousands of files, it is practically impossible to maintain the correct information on relationships between files.

Information on the procedures used to develop a product needs to be related to the product and to the activities. Similarly, information on the procedures that should be followed for new products should be linked to the activities and the files. In the absence of links between the information and the process that creates it, it is impossible to get real control of the environment. This is true from the technical point of view where information concerning relationships or selection criteria for alternatives is lost. It is equally true from the management point of view, with the result that the engineering function does not operate as efficiently as it could.

5.9 MEANING OF DATA

Raw data often has little meaning and is of limited use, yet, in the traditional engineering environment, it is often only raw data that is stored. Typically, the data that is maintained on a product will include CAD/CAM files, some drawings, and some information on the manufacturing process. Little information will be maintained on user needs, the process that was used to develop the data, or the choices that were made to get to this data. Most of the information will be forgotten. The next time similar activities are carried out, it will not be possible to benefit fully from past experience, and development will start again from scratch.

Drawings and CAD/CAM files are both examples of raw data. They do not provide complete definitions of the product, the process used to develop the product, or the various choices and activities that took place during the design cycle.

A lot of information is discarded during the design cycle. For a particular phase of the design cycle, this may not be a major problem, since the people involved can remember what has been discarded. It becomes a problem, however, when the data is transferred to other users in another part of the design cycle, since they will not be aware of what has been discarded. It is also a problem when an attempt is made to reuse information at a later date. By this time, the original users of the data will have forgotten what they discarded. Time will be lost as they try to find missing data or to develop new data to replace missing information.

Unless users understand the full meaning of data, they cannot make the best use of it. The information that could be of use to them needs to be available to them. With current systems, this is rarely the case. In the typical engineering environment, a lot of the raw data that is created is soon discarded because there is little chance of being able to use it profitably again. Users need support from management systems so that they will not discard potentially useful data.

Knowledge-based systems could be used in this role. Knowledge-based systems may also be used to help "renovate" existing, incomplete data. The company's existing data is valuable, but, to unlock its value, users have to know its meaning. Unless the full meaning of data is available, it cannot be fully used in the future. If data has to be re-created or renovated manually, errors will be introduced and time will be wasted.

5.10 LIFE CYCLE DATA

Engineering data supports the product life cycle. In some cases, for products such as aircraft, the overall product life cycle may be 30 or 40 years. During this time, there will be a vast volume of data generated, first to design and manufacture the product, and then to support its use.

Generally, the support cycle is longer than the design and manufacturing cycle, and may produce correspondingly more data. Technical manuals will have to be produced and kept up to date. Spare parts will have to be ordered and manufactured. For each maintenance job, information will need to be available on the handling tools and the required skills. Product specification data may need to be given to second sources. Data supporting repair and replacement will be needed. Field data needs to be managed. Performance data needs to be maintained, so as to be able to plan preventive maintenance.

A lot of data will be produced during the product life cycle. Many users, perhaps in different companies, will want to access the data. Many different activities will make use of the data. Each will want the data to be available in the most suitable place and format. Different types of data will be produced and needed at different times. New data will be produced, existing data will be reused and perhaps modified. Over a long life cycle, the product may be repaired or upgraded to such an extent that most of the original product will have been replaced. At all times, the configuration documentation must correspond to the state of the product.

An aircraft may need to be repaired in any part of the world. Information must be available on its exact configuration at any moment and in any location. Similarly, product configuration data may be required anywhere in the world for naval vessels. If a ship has problems in the Antarctic, it is preferable to know the exact configuration and to be able to fly a spare part out to it, rather than bringing it back to a Northern Hemisphere port to find out which parts are currently on board.

At any moment, it may be necessary, for any of a variety of reasons, to look back at the design of a particular part or batch. A batch of biscuits may be inedible. A batch of airbags may be faulty. A part may have failed on a 30-year-old airplane. Many companies must keep original drawings of their products, going back over many decades. They must be able, for legal reasons, to trace back the components of their products.

In the future, companies will have to keep data that is on electronic media for similar periods and similar purposes. In the same way that some traditional media deteriorate, some electronic media will also not be suitable for long-term storage.

Access to original electronic data will be complicated by the rapid obsolescence of computers and the introduction of radically different systems. New content-rich CAD/CAM systems will pose a problem. They will have difficulty in using old 2-dimensional data produced for old products by old systems. Just as geometric modelling programs used for aircraft design in the mid-1960s have been super-seded, today's programs will, in turn, be superseded. Companies will be faced with keeping old data (for old products), enriched data on old products (since, without some renovation, the data might not be useful), and data in new formats.

An audit trail needs to be kept so that it is possible to go back in time and see how and why a particular part was made. An audit trail is essential in locating and correcting design errors. Security needs to be maintained throughout the life cycle. It is not enough to maintain a secure environment during the design phase. Information must also be secure during the support cycle. Just as some designers will not have the right to see some information, some maintenance staff will also have limited access rights.

During a long product cycle, major problems can be caused by the departure of key individuals. In some specific areas, they may have unique knowledge, and, unless the necessary actions are taken, their departure can have unforeseen effects. This is another activity for which knowledge-based systems will be useful.

5.11 A COMPLEX BUT LOGICAL ENVIRONMENT

A large volume of engineering information of various types is used by a variety of people from different organizations working at many activities. There is consider-able overlap in the use of information between different individuals and organiza-tions. There is also an overlap of activities between people, functions, and compa-nies. Products, processes, technologies, and philosophies undergo continual change. This is a complex environment. Until now it has been left to run itself. However, for the reasons described above, it is becoming increasingly clear that this environment must be managed better.

On close examination, though, it becomes clear that, even if the environment is complex, it is also logical. The process of data creation and use, release, change revision, and archival is complex but straightforward. It follows a set of logical steps. As such, there is no reason why it should not be successfully automated. If it is not, and the number and complexity of processes and variants continues to increase, manual systems will eventually break down. Similarly, the process of distributing and transferring data through the company and to suppliers, partners, and customers is basically straightforward. It becomes complex when it is not controlled. If it is felt that the environment is impossible to control, then it will be impossible to control. On the other hand, once it is accepted that improvements can be made to the environment, it becomes possible to make improvements.

In most companies, the current flow of engineering information and processes is unnecessarily sequential and time-consuming. As a first step to reducing lead times, every activity in the engineering process should be identified and examined both from the point of view of the individual activity and from the point of view of the other activities in the process. This task should be carried out by a multidisciplinary team. It will be found that many activities can be profitably redefined or even eliminated. To reduce lead times further, systems can be put in place to support overlapping and parallel activities, and partial release.

In the same way that product design can sometimes be parameterized by techniques such as group technology, once the design process is clearly understood and managed, it too can be parameterized. It will then become possible to design new products much faster.

Information on the procedures to be used in an activity can be associated with the activity. Conventional project management tools can be integrated with the management of engineering information and activities.

Rather than just managing CAD/CAM data files, EDM/EWM systems can play a prime role in improving engineering performance, significantly reducing lead times and improving product quality.

5.12 SUMMARY

The increasing pressures on business stemming from a more competitive, global market lead to the need to increase quality, reduce cost, and reduce cycles. At the same time, they also lead to an increased complexity of doing business. These global factors filter down directly from the level of top management and corporate strategy to the engineering function and engineering management. At this level they become very clear targets.

Lead times must be reduced. Products must be brought to market faster. As product lifetimes get shorter, significant market share is lost if a product is not brought to market at the earliest possible moment. A company that gets to market first can capitalize on late market entry by other companies. Product costs must be trimmed so that they correspond exactly to customer requirements. Similarly, product functionality must be improved to match these requirements.

Engineering managers know only too well how difficult it is to meet these challenges. Too often, they find that, in spite of all their efforts, they finish up with the same results of late delivery and poor quality. They know that sometimes they are not in complete control of the projects entrusted to them. Yet, as they look round, they see the potential for improving the situation. They see time wasted, as drawings are manually transported from one activity to another. They see time wasted as information waits to be signed off. They see an unnecessarily high number of signatures holding progress back. They see manufacturing personnel doing nothing, as they wait for information to be released from engineering. They are aware that technical manuals are often obsolete. They are aware of the scrap, the personnel time that is wasted, and the possibilities of reducing the design cycle. They know that late market entry is going to cost the company a lot of money. They do not set out to miss the targets that top management gives them, but they often misjudge the technological, organizational, and human barriers that prevent them hitting those targets. They understand as well as everyone else that the time to reduce production costs is during the engineering phase, when the production costs are defined. They understand that it is much cheaper to reuse or slightly modify an existing design than to start a new design from scratch. Like everyone else, they know that it is important to control engineering changes and maintain exact product configurations. Unlike everyone else, though, they are responsible for getting these things done, and in practice it is not as easy as it sounds.

Engineering managers need the tools to help them manage the engineering workflow and engineering information. Then they will be able to meet the targets. An EDM/EWM system is one of the most important of these tools.

Any system that is put in place to manage the engineering workflow and engineering information must be sufficiently powerful to maintain control, yet flexible enough to allow the changes that typify the engineering environment. The system needs to be able, for example, to allow partial or early release of data. It needs to allow changes to be made as work progresses. It needs to be able to manage activities that are still primarily paper-based, as well as those that are CAD/CAM-based. It needs to be able to handle scanned paper drawings, as well as the data in CAD/CAM files. It needs to be able to work with different levels of data definition and with different representations of data.

Currently available systems do not meet all the requirements for management of the engineering process and engineering information. Until the introduction of EDM/EWM systems, the two major contenders were probably CAD/CAM systems and the data bases found throughout the financial and commercial world. However, the traditional CAD/CAM system is focussed on providing the design, analysis, drafting, and machining functions necessary for its users. Its developers have not attempted to manage the whole engineering environment. They did not set out to improve workflow or to manage engineering data not used by the CAD/CAM system. Similarly, traditional commercial data bases were developed to meet other requirements than those stemming from the engineering environment.

As existing systems did not fully meet the requirements of the engineering environment, a new type of system—the EDM/EWM system—was needed.

6

Data Base Management Systems

6.1 DATA IN FILES AND DATA IN A DBMS

Having introduced the concept and components of an EDM/EWM system, and described the need for an EDM/EWM system from a business point of view, this chapter and the two following chapters describe EDM/EWM issues from the data viewpoint. Some readers may now prefer to read Chapters 9 and 10, describing EDM/EWM implementation issues, and then come back to the detailed data issues.

As one of the major reasons for the increasing interest in EDM/EWM systems is the ever-growing volume of engineering data, one apparent solution would be to put this data in a data base of the form so commonly used for financial and other commercial applications. To explain why such a solution is not suitable, a brief introduction to some data base concepts is needed.

Computer programs use operating system functions to store data in files. In a very small organization with few contacts to the outside world, it may be possible to use an essentially directory/file-based data management system (Figure 6.1). Different parts of the file name can be given specific meanings. One part of the file name can represent a project name, another can represent a version number, and a third can represent the release status. A library of frequently-used parts can be built up using similar techniques. The operating system's password structure and privilege rights can be tailored to restrict access to authorized users. Archiving and backups can be handled automatically. For some time, it will be possible to work like this. However, users will eventually grow tired of file names like truck7_roof43_front2_right5_try7.ds8. They will be handicapped by their lack of ability to add extra information to the file name. They will not want to share data with other users, as they will be worried about data corruption. Even in a small organization, it does not take long to reach the limits of file-based data management techniques.

```
Directory: truck.glass43

Glass1.model4.xl1.ds8    1940    12-07-82   ww   r
Glass1.model4.xl2.ds8    2010    12-08-82   ww   r
Glass1.model4.xl3.ds8    2218    12-11-82   ww   r
Glass1.model4.xl4.ds8    2034    12-12-82   ww   r

Directory: truck.glass44

Glass1.model1.ds8         920    12-12-82   ww   rw
Glass1.model2.ds8         950    12-12-82   ww   rw
Glass1.model3.ds8        1088    12-13-82   ww   rw
Glass1.model4.ds8        1046    12-14-82   ww   rw

Directory: truck.roof43

Abeam1.right1.try1.ds8    245    01-01-83   hk   r
Abeam1.right1.try2.ds8    235    01-05-83   hk   rw
Abeam1.right1.try3.ds8    254    01-05-83   hk   r
Abeam1.right1.try4.ds8    253    01-05-83   hl   rw
Abeam1.right1.try5.ds8    253    01-04-83   hk   r
Abeam1.right1.try5.ds8    234    01-03-83   hk   rw
Abeam1.right1.try7.ds8    244    01-08-83   hk   r
Abeam1.right1.try8.ds8    254    01-05-83   hk   rw
Front1.right1.try1.ds8    483    01-01-83   rj   r
Front1.right1.try2.ds8    485    01-01-83   rj   r
Front1.right1.try3.ds8    486    01-02-83   rj   r
Front1.right1.try4.ds8    487    01-03-83   rj   r
Front1.right1.try5.ds8    483    01-04-83   rj   rw
Front1.right1.try5.ds8    495    01-03-83   rj   r
Front1.right1.try7.ds8    496    01-03-83   rj   r
Front1.right1.try8.ds8    489    01-03-83   rj   rw
Front1.right1.try9.ds8    476    01-04-83   rj   r
Front1.right1.trya.ds8    509    01-05-83   rj   r
Glass1.bills1.ds8        2034    12-12-82   ww   r
Blend1.my1.try1.ds8       532    01-10-83   rj   rw
Blend1.my1.try2.ds8       603    01-12-83   rj   rw
Blend1.my1.try3.ds8       582    01-12-83   rj   rw
Blend1.my1.try4.ds8       643    01-15-83   rj   rw
Blend1.my1.try5.ds8       552    01-15-83   rj   rw
Blend1.my1.try6.ds8       623    01-12-83   rj   rw
Blend1.my1.try7.ds8       562    01-12-83   rj   rw
Blend1.my1.try8.ds8       673    01-11-83   rj   rw
Blend1.my1.try9.ds8       663    01-12-83   rj   rw
Blend1.my1.trya.ds8       662    01-12-83   rj   rw
Blend1.my1.tryb.ds8       623    01-16-83   rj   rw
```

FIGURE 6.1   A directory/file-based data management system

In a large organization, it is almost impossible to manage file-based data efficiently, yet most of today's engineering computer systems are file-based islands of automation. They suffer from several disadvantages. Users lose track of their data in a sea of distributed files. It is difficult to share data between systems. There is a lot of data redundancy, as duplicate items of information are held in several systems. It is difficult to control access and to maintain security of data in several independent file-based systems. It is difficult to apply uniform rules and standards to data that are held in several file-based systems, especially as a lot of the data logic may actually be in the programs and not with the data in the files. As the data are not completely independent of the program, any changes to the program are likely to affect the physical and logical structure of the data. The result of this is that an enormous amount of time and money is spent in maintaining existing programs, thus holding up new developments.

The DBMS (Data Base Management System) approach sets out to overcome these problems and to manage data as an important company resource. It takes a top-down, global approach to data and is oriented towards data, rather than towards individual programs. It aims to provide multiple users with efficient, secure, and convenient access to data.

A data base is a collection of interrelated information stored with minimum redundancy, usable by several programs, but independent of these programs. A data base management system (DBMS) is software that manages the data base. Users are only allowed to access the data base through the DBMS. Among the major elements of a DBMS are the data model, the data dictionary (DD), the data definition language (DDL), the data manipulation language (DML), data management functions, and query programs (Figure 6.2). These are described in more detail below.

The DBMS approach aims to :

- Provide for physical and logical data independence,
- Provide for explicit data validation rules and standards,
- Provide for controlled access, security, and recovery,
- Reduce data redundancy, and
- Allow data to be shared by many different users with different requirements.

A simple example from the commercial environment will illustrate some of these points. Many of the data models in the commercial environment can be thought of as having a two-dimensional structure of records and fields. Information on a given item, such as a customer, is stored in a record. Each record is divided up into fields (Figure 6.3). For a customer record, the fields could hold information on items such as name, address, telephone number, product purchased, purchase date, and credit rating.

Without a DBMS, all users of this information would have used programs that had their own associated files of data (Figure 6.4). Each of these would have had a different structure of records and fields. One program might have had one record per customer; another, one record per product; and a third, one record per purchase. Each program would store the data it required in fields. As some of the information, such as customer names, would be needed in several programs, there would be redundant data. Information such as customer addresses would also be needed in several programs (e.g., invoicing, mailing list, delivery routing) and be repeated in several files.

When data needed to be updated, there would always be the risk that some references would not be updated. A change of customer address might be entered into the mailing list file, but not on the delivery routing file. Purchase orders,

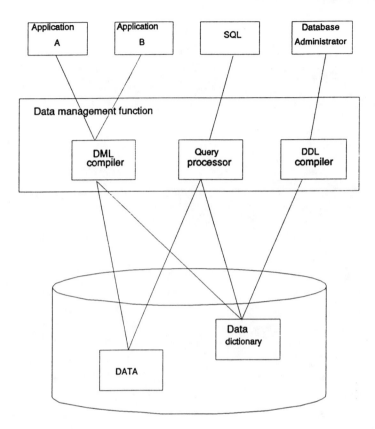

FIGURE 6.2   Major elements of a DBMS

payments received, and credit ratings might be in files controlled by different programs, with the result that, if these were not all updated properly, an order might be turned down because the most up-to-date payment information was not available to the credit control program.

Implementation of a DBMS overcomes these problems and provides benefits such as a common repository for data, elimination of redundant data, improved security, and access control (Figure 6.5). All users of the data, from whichever department, make use of programs that go through the DBMS to add, view, or use data. As soon as information, such as a change of customer address or the receipt of a payment, is in the system, it is available to all users. A clear parallel can be seen with the situation in engineering, where there are numerous uncoordinated programs and data files. It has been proposed that the DBMS found in the commercial environment be used in the engineering environment.

Four customer description records of five fields

Smith	Queen	313-1234	7	Tuesday
Jones	Main	212-1234	4	Monday
Black	East	703-1234	9	Friday
Green	North	405-1234	6	Thursday

Animal description records

Cat	Black	4	Tulsa
Dog	Tan	4	Menlo
Bird	Blue	2	

Purchase records

53	Cat	Girl	Maui	Jan 27	Hardy
54	Cat	Boy	Tulsa	Jan 27	Smith
55	Dog	Lady	Menlo	Jan 28	O'Hara

FIGURE 6.3   Records and fields

6.2 DISADVANTAGES OF COMMERCIAL DBMS IN THE ENGINEERING ENVIRONMENT

There are major differences between engineering data bases and commercial data bases of the type commonly used for payroll data, airline reservation data, and bank customer data. Commercial data bases will typically contain only alphanumeric information. They do not handle the mixed data types of the engineering environment. They rarely handle graphic data.

In commercial data bases, all records relating to a particular item (such as a customer) have similar lengths. In the engineering environment, data on two products of the same type (such as two gear wheels) could be in the form of a different number of records of different record lengths. The data structure of the two products could also be different, reflecting different product structures.

In the commercial data base environment, a transaction, such as a seat reservation,

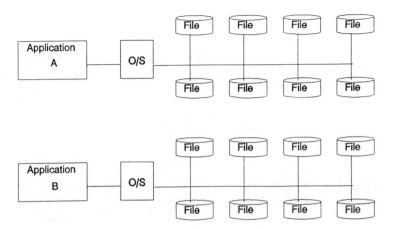

FIGURE 6.4   A file-based approach

typically lasts a few seconds or minutes. In the engineering environment, a transaction, such as the design of a new part, will generally last several hours or days.

In the commercial data environment, the applications are built round the DBMS. New data is then added to the DBMS. The majority of data should therefore be in the DBMS. In the engineering environment, a lot of the data already exists and is not in a DBMS. Most of the major creators of data, such as CAD programs, do not feed their data into a DBMS, but into separate files. Some of the records in the engineering environment are so long that they cannot be handled by commercial DBMS.

In the commercial data environment, the relationships between data are generally static, well specified, and simple. In the engineering environment, they are dynamic, unclear, and complex. The engineering environment is characterized by

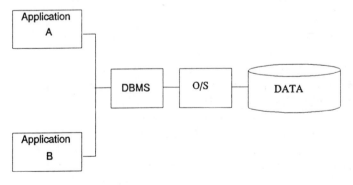

FIGURE 6.5   The DBMS approach

versions and alternatives of the same basic part or product. These occur less frequently in the commercial environment.

In the commercial environment, modification to one record will not normally require modification to a large number of other records. In the engineering environment, modification to one record, say a part design, could lead to a large number of related modifications. A typical commercial DBMS does not address the workflow considerations that arise in the engineering environment.

In view of the above differences, there are few engineering organizations that can make use of the type of DBMS that is used in the commercial environment, for more than a small fraction of the engineering data. However, in view of the obvious disadvantages of file-based systems and the potential advantages of data base management systems, engineering companies recognize that a DBMS approach is needed. In response to this demand, engineering-oriented DBMS solutions are becoming available.

6.3 METADATA

For various reasons (in particular, the great differences between the different sets of data to be stored), the solution of putting all engineering data in a single DBMS is generally rejected. An alternative approach, lying between the two extremes of a purely file-based, non-DBMS solution and a full DBMS acting as a repository for all data, involves "metadata." This approach involves leaving the data in the files in which they were created, and only putting in the DBMS a limited amount of information about each file. This "data about data" is often referred to as metadata or directory data (Figure 6.6).

It is the metadata, rather than the underlying data, that is stored in the DBMS. The information stored can include file location, file name, program creating, creator, owner, status, creation date, release date, release procedure, reviewer,

Part number
Part name
Version
File name
Drawing number
Owner
Access rights
Related files

FIGURE 6.6   A metadata record

releaser, part name, data type, data structure, data format, rights, version number, promotion level, project, references, log information, and so on. Other fields in the metadata record could refer to specific attributes such as material, size of fasteners, and so on. The files can be maintained in their native format in their usual locations. The system can automatically follow what is happening to the file, automatically writing the metadata and alleviating the user of this work. The metadata can be extended to include everything that needs to be known about the information in the file and the way that it fits into the engineering process.

Users can request searches of the metadata records, to find specific information. A user may want to find where a given file is stored. It may be that there is no corresponding file, but a drawing on paper, in which case the metadata could point to the physical location of the drawing. Once a file is found, the system can retrieve it and modify its access status (e.g., to view-only).

The metadata record is similar in appearance to a record in a commercial DBMS. It is of an equivalently normal and standard length. It only contains alphanumeric characters. It can contain information on the relationships to other files. The number of metadata records will be of the same order as the number found in a commercial DBMS. The metadata records can therefore lie at the heart of an engineering data management system.

The functionality of an EDM/EWM system will of course be specific to the engineering environment. However, many of the data base components of the EDM/EWM system will be the same as those in a commercial DBMS.

6.4 DATA BASE VOCABULARY

Some users of the data base will need to have a detailed understanding of the way data is physically stored. Other users may only need to know about, for example, the current value of a particular field. To obviate the need for all users to work at the most detailed level, a multi-level approach is used (Figure 6.7).

The lowest level is the physical level. Here, low-level data structures are described in detail. This level describes how data are actually accessed and stored (e.g., on a particular disk and with a particular layout on the disk). The typical user does not want to work at this level, but at a higher, logical level, at which it is possible to interact with individual attributes of parts or with complete models of objects.

Data models are used to describe the structure of a data base. The data model describes the organization of relationships between items of data, as well as the constraints on data. There are several different data models in use. Physical data models describe data at the lowest level.

At the next level up, the logical level, a description is given of the data actually stored in the data base, as well as the relationships between them. This level is

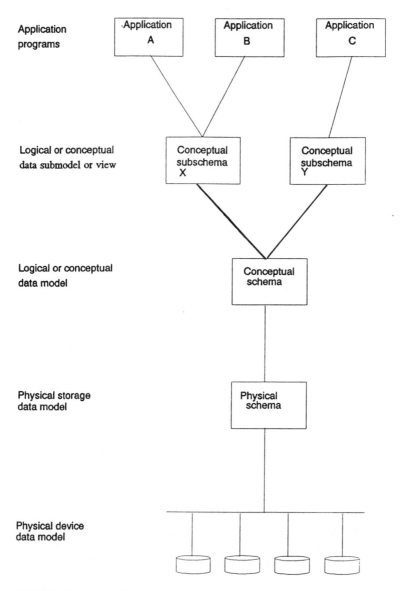

FIGURE 6.7   The multi-level approach

sometimes referred to as the conceptual level. A logical data model describes the overall, global view of the organization's data. This description is independent of the actual way that data is physically stored. Record-based logical models and object-based logical models describe data at the conceptual level. The most commonly used record-based logical models are the hierarchical model, the network model, the index model, and the relational model. They are described in more detail below. Among the most appropriate object-based logical models for the engineering environment are the entity-relationship model and the object-oriented model.

As most users do not want to access all of the information in the data base, they do not need to be aware of all of the logical level description. They only need to work with a certain well-defined part of the data base called a "view." They are provided, at the view level, with a conceptual description of that part of the data base that is of interest to them. This is sometimes called a logical data sub-model.

The overall design of the data base is called the data base schema. Corresponding to the levels described above, there will be a physical schema, a conceptual (logical) schema, and subschemas corresponding to views. The logical schema contains the names of record types and fields, and their relationships. Constraints can be applied in the schema to control access and to maintain security and integrity.

A schema is defined using the data definition language (DDL). This is used to define data elements and the relationships between them. The data definitions are compiled to give a set of tables that will be stored in the data dictionary (DD).

The data dictionary maintains data definitions and the rules governing data access. The DD contains the schemas and subschemas (logical structure) of the data base, describing the view that a particular program or user has of the data. The DD maintains information on the location of a particular data element, as well as corresponding access and retrieval methods.

The data manipulation language (DML) allows access by a computer program to the underlying data. The DML allows users to access, insert, and manipulate data. With a procedural DML, a user specifies what data is needed and how to get it. The DML commands are embedded, for example, as a subroutine call in a program written in a host language. With a nonprocedural DML, a user just specifies the data to get.

Most DBMSs offer query languages that are easy to learn and easy to use, and can be used independently of computer programs to allow casual users direct access to data. Generally, they also include report generators that allow users to easily describe the format and contents of screen layouts and reports.

6.5 TRADITIONAL DATA MODELS

The three most common data models are the hierarchical, the network, and the relational data model.

In data base terminology, an entity is something, such as a product, a part, a machine, or a customer, about which the company will maintain data. In the hierarchical model, entities are organized in a tree, or hierarchical structure (Figure 6.8).

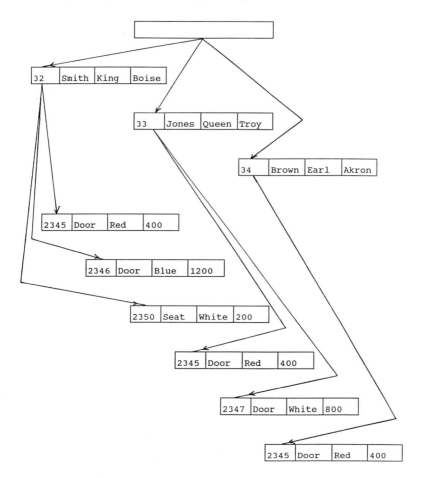

FIGURE 6.8   The hierarchical model

The uppermost level of the hierarchy, the "root," has no "parent" above it. At all other levels, entities have one parent. Each entity can have one or more "children" at the level below it. An example of such a structure is a planet composed of continents, each composed of countries, cities, each composed of streets, and buildings. In the hierarchical model, entities are implemented as records. Each record is made up of a collection of fields. A field of data is an attribute (e.g., address of a customer) of the entity. Records are connected together through links, or pointers. Only one hierarchical structure can exist in the model. It is defined by the data base designer. Since data can be organized in the way that it will be used, the hierarchical model can provide very high performance. However, when an attempt is made to access data other than by the chosen path, performance can be very poor. If the access sequence is from parent to child, the model is efficient. If, however, it is from child to parent, or from child to child, it can be very inefficient. Rearrangement of the structure is not easy, and may involve changing each individual record.

The network data model (Figure 6.9) is similar to the hierarchical data model, except that it allows entities to have many parents at the level above them, while still having one or more children at the level below. Entities are represented by records. Each record is made up of fields. Records are connected by links.

Network and hierarchical data bases can be tuned to become very efficient for specific applications. However, this requires considerable expertise and the introduction of redundant and perhaps inconsistent data and data access paths. Perhaps

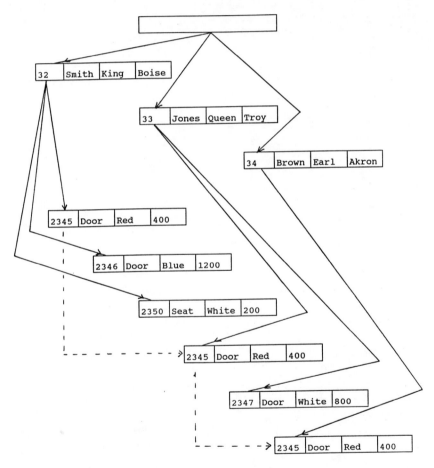

FIGURE 6.9   The network model

most importantly though, their major disadvantage is that the application programs that use them need to be aware of the data base structure. If the structure changes, all the application programs may need to be changed.

In the index, or inverted file data, model, structures are similar to the hierarchical and network structures, except that instead of using embedded pointers to maintain relationships, this information is stored in separate tables. However, there is still a physical orientation to data, and users need to be aware of the data base structure.

In the relational data model (Figure 6.10), data are represented in two-dimensional

The supplier relation

Supplier Number	Supplier Name	Supplier Address	Supplier City
32	Smith	King	Boise
33	Jones	Queen	Troy
34	Brown	Earl	Akron

The part relation

Part Number	Part Name	Part Color
2345	Door	Red
2346	Door	Blue
2347	Door	White
2348	Seat	Red
2349	Seat	Blue
2350	Seat	White

The order relation

Supplier Number	Part Number	Quantity
32	2345	400
32	2346	1200
32	2350	200
33	2345	400
33	2347	800
34	2345	400
34	2348	400
34	2349	200
34	2350	1000

FIGURE 6.10   The relational model

tables known as relations. Each table (relation) is made up of horizontal rows (tuples) and vertical columns (attributes). The number of rows in a relation is the cardinality of the relation. The number of columns in a relation is the degree of the relation. From the user's point of view, there is no hierarchy in or between tables—the tables are "flat." Each table has a distinct name. Each row is unique, and each column has a distinct name. Rows are associated by common column entries. The values in a column (i.e., of an attribute) are drawn from a domain of allowed values. The data dictionary stores information about the relations (e.g., names of relations and attributes, names and definitions of views).

One result of the relational data structure is that it allows application programs to become independent of data. The program tells the DBMS what data it needs, and the DBMS finds the data. As a result of the structure of the relational model, data can be added or deleted from the data base without the need to change application programs. New tables can be added, and new rows and columns added to existing tables, without reorganizing the existing data base. New tables can be produced by combining existing tables, under the rules of the mathematical theory of relations known as first-order predicate logic. The data independence offered by the relational data model cuts the workload and increases the flexibility of programmers. Another major advantage is that little or no knowledge of detailed data structure is required by casual users.

Relational data bases offer great flexibility, but can suffer from poor performance when large volumes of data are stored. Individual operations are generally much slower than if programmed in a standard programming language. Relational data bases using the common query language SQL are limited to one-dimensional indexes. This makes searches in large data bases very slow. Searches go much faster with the type of order-preserving structures associated with hierarchical data bases.

In the complex environment of engineering data, relational data bases also suffer from only having simple data types and structures. Although relational data bases are not currently the solution for all the engineering data, they do represent an appropriate technology for management of engineering metadata.

Enhancements are being made to some relational data bases to handle new data types, to offer custom indexing, to allow sharing by multiple data bases, and to provide built-in forward- and backward-chaining. In time, the extended relational data base may become a suitable technology for all of a company's engineering data.

6.6 DEVELOPMENTS TO TRADITIONAL DATA MODELS

Some of the problems of relational data bases can be overcome through enhancements. For example, to overcome problems associated with handling very long data items, a file system can be embedded in the DBMS. Special data types can be included to cater for hierarchical relationships. More use can be made of object-based methods. In the relational model, the data is stored in flat tables, but the relationship between data has to be maintained by the application program. In the entity-relationship model, an object-based logical model, the relationships are considered to be "data," rather than part of the application.

An entity is a basic object that is distinguishable from other objects and about which data can be stored. A set of descriptive attributes is associated with each object. A relationship is an association among entities. An entity set is a set of entities of the same type (e.g., the set of all employees in the engineering department). A relationship set is a set of relationships of the same type. Aggregation allows relationship sets and associated entity sets to be treated as a higher level entity set that can be treated in the same way as other entity sets.

The model can be expressed graphically in an entity-relationship diagram (Figure 6.11), in which entity sets are represented by rectangles, attributes by ellipses, and relationships by diamonds. Entity sets, attributes, and relationships are linked by lines.

Traditionally, the entity-relationship model has been used for the task of data base design, but not as an implementation structure. However, data bases are now appearing that use it in this way.

An object-oriented data base contains objects defined by an object-oriented data model. In an object-oriented environment, each piece of information is defined as an object. Objects are defined by an object-oriented data model that covers objects, attributes, constraints, and relationships. Each object is an instance of a class of similar objects with common characteristics. Associated with each class is a description of common methods and procedures for building and handling objects within that class. The data and the methods are stored together. This technique, of

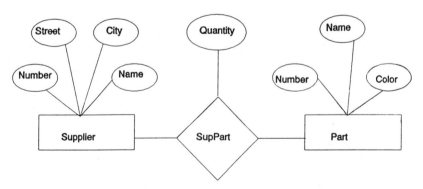

FIGURE 6.11   An entity-relationship diagram

storing both information describing objects and the methods or processes that work on the objects, is known as encapsulation. Users can only work with an object through its methods. A message is sent to the object capsule, requesting that a particular encapsulated procedure be applied to the object.

New objects can be created from existing objects, inheriting all the characteristics of the existing object. Compared to relational data bases, object-oriented data bases offer new features of encapsulation, inheritance, and polymorphism—the ability to uniformly apply processes to a variety of objects.

As the description of an object also includes a description of the process and relations between objects, it offers possibilities for artificial intelligence applications. Specific know-how can be stored with the object.

The object-oriented approach is attractive in the engineering environment where "objects," such as a product, a part, a drawing, and a manual, are natural units of work. However, object-oriented data base technology is still in its infancy. In view of the amount of time that was needed before data base systems based on traditional models were available for widespread, everyday use, it can be assumed that it will be many years before object-oriented data bases play a significant role in the engineering environment.

6.7 FROM DATA BASES TO EDM/EWM

Data base technology is an important, but relatively small, part of the overall EDM/EWM solution. Organization of the engineering data and organization of the engineering workflow are equally important issues.

Questions will arise as to whether the data and/or the metadata should be centralized, and as to how much of the metadata should be distributed. The answers to these questions will depend on the way a given company has been and intends to be organized, as well as the benefits it intends to gain from EDM/EWM. They will depend on the volumes and types of data, and on security, integrity, and performance requirements.

Data could be logically and physically centralized. It could be logically centralized, but physically distributed. It could be both physically and logically distributed. Metadata could also be physically or logically distributed or centralized. The system could be such that the user may be aware or unaware of the location of data. If the data is distributed, it could be fully or partially connected in a tree, star, or ring network. For recovery purposes, at least one copy of data may be kept. This could also be centralized or distributed. Copies may also be used to reduce communication overhead. The system could be such that the user may be aware or unaware of the existence of copies of data. If data is distributed, synchronization of updates needs to be addressed.

A company could put all of its data in one data base. Alternatively, it may only have the data for a particular function or group in the data base. Data that is used by many users may be stored in a centralized data base, while data specific to an individual user or to a group of users may be stored locally. A company might make use of a "database machine"—a computer system with hardware and software optimized to handle high-performance data handling functions.

A company could approach EDM/EWM through a pilot phase, step by step, or through a "Big Bang." It is not sufficient just to decide what to do with the data. The rights of users and the degree of local autonomy for decentralized users will need to be defined. It will have to be decided how to enforce the rules and how to maintain semantic integrity across distributed data bases. In a distributed environment, it will have to be decided how data will be fragmented between different locations, and what level of communications load is acceptable.

Development of a data base requires considerable organizational investment in time and money. Who will pay for this? Will those who pay automatically control the system and own the data? What will happen to existing files—will they all have to be converted? How can the potential benefits of EDM be financially justified?

These questions and many others like them relate much more to organizational issues than to pure data base technology. Unless suitable answers are found, even the best technological approach is unlikely to succeed.

7

Modelling Engineering Processes and Information

7.1 THE NEED FOR MODELLING

A prerequisite for effective use of an EDM/EWM system is a complete and clear description of engineering information and the associated processes. Simple models of this complex environment must be built. Models will be needed to describe both existing and planned systems. Models will be needed to describe information, to describe processes, and to describe their interaction. They will serve as a formal basis for the development and implementation of systems. A model may be built of several submodels. It will probably be in the form of text and graphics. It should clearly represent the dynamic and static relationships between information, processes, events, and controls.

Development of models should involve many people in the company. Involvement in this activity helps them better understand the work they do and the way that it fits into the overall workings of the company. Their involvement at this stage will increase their commitment to the eventual use of the systems that are developed. The model acts as a common basis for communication. It offers the opportunity for getting agreement among people who previously had different views and definitions of the environment. It will help people gain an understanding and reach a common view of the situation. Development of the model will help to get people to think cross-functionally. The development of a model is generally an iterative approach. The first attempt at a model will probably lack detail and be incorrect, but it provides a starting point from which further refinement can take place (Figure 7.1).

The structure and use of information and activities within the engineering

97

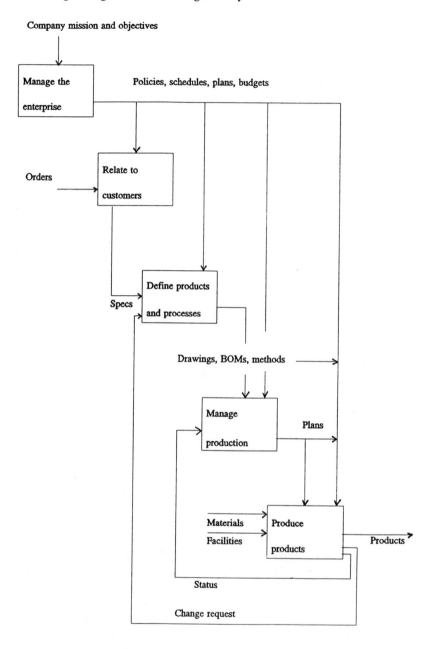

FIGURE 7.1   An initial model

environment must be understood. The development of a model helps visualize, clarify, and complete their definition. Modelling will help people understand the use and flow of engineering information and the engineering process. The relationships between activities, systems, and people will become clearer. It will be possible to understand the events that link them and to identify the major management milestones used to control them. The modelling activity can be addressed in several ways, many of which are complementary. It is useful to build up pictures both of the current ("as-is") situation and the future ("to-be") situation. These will eventually be related by the EDM/EWM development plan. Another useful pair of models is the "top-down" model and the "bottom-up" model. These should of course agree in the middle. The "top-down" model is derived from a business-oriented description of the engineering process, working down towards individual operations and detailed descriptions of data and activities. The complementary "bottom-up" approach starts from individual operations and detailed descriptions of data and activities, and then links data and operations and builds successively higher levels of information and processes.

In a similar way, the physical implementation of the engineering activities and the paper (or computer-based) model of the engineering environment that is developed are complementary representations. The paper model will be used as the basis for developing the EDM/EWM environment, first as a logical representation and then as a physical computer-based representation. The models must take account of both the information and control flow, and the information and control structure.

Various methods and diagramming techniques have been developed to support the system development modelling activity. Among the most commonly used are those developed by Yourdon/DeMarco, Warnier/Orr, and Nassi/Schneidermann.

7.2 FOUR TYPES OF MODEL

The requirement to understand both information flow and structure, and control flow and structure, leads to four distinct types of models for data flow, control flow, data structure, and control structure.

Control-structure models focus on top-down decomposition of the control organization. One example is the typical organization chart showing the various departments of a company, the groups within the departments, the teams within the groups, and the individuals within the teams (Figure 7.2). Another example of a control structure model can be drawn from the organization of a computer program and its use of subroutines. Another example, specific to the engineering environment, is that of the hierarchical structure involved in sign-off and release of design work.

Control-flow models describe the allowed values and combinations for the inputs and outputs of functions, and show how these are related to the detailed activities of the company (Figure 7.3). An example of a control-flow model is the

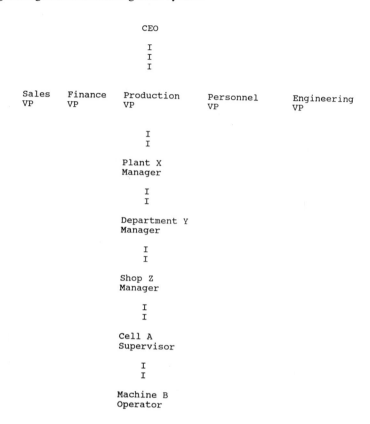

FIGURE 7.2   A control-structure model

PetriNet. A programmer's flow chart is also an example of a control-flow model. It shows the exact sequence of instructions to be executed and the points where flow is dependent on the value of particular variables.

Data-flow modelling, which will be described in more detail later, shows how data flows through a system, and by which activities it is processed and stored. For each activity, the information that is input and output, as well as reference and control information, has to be understood. Data-structure modelling, which will also be described in more detail later, focuses on the structure of data elements and the links between them.

7.3 THE APPROACH TO MODELLING

Before describing some of the models in more detail, it is useful to consider the environment in which modelling is likely to take place. For several reasons, it is

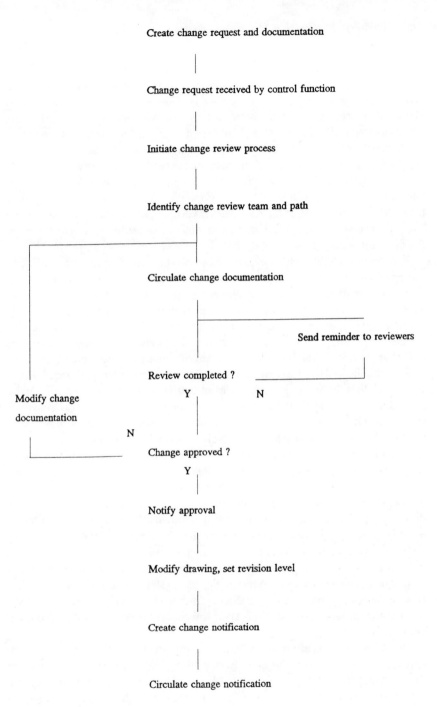

FIGURE 7.3 A control-flow model

likely to be conflictual. Individual engineering personnel will probably only be interested in a better system for managing their own data. They would like to see this system available immediately and without any effort on their part. They will not be interested in the overall information process, nor will they be interested in the procedure that is taken to develop a better engineering data management system.

On the other hand, some of the other people involved in defining the new system will have a different, if not opposite, approach. They will be specialists in developing the logical design of the overall system and of the detailed data base. They may well have no interest at all in the information itself or in its use. They may have little understanding of the real issues involved in engineering, and, following a particular methodology, generate vast volumes of superfluous data that obscure the most important needs.

Parallel to the different behavior of these two groups of people may well be a disagreement over the best way to develop the new system. The engineers, looking for a quick solution to their own problems, will appreciate a prototyping approach. This is an iterative approach that accepts that user requirements cannot be clearly defined initially, and in any case will change as the users get to know the new system. A "prototype" is built to meet the user's apparent requirements. Experience of its use provides input for the development of a more advanced prototype.

The classical EDP approach to system development, often preferred by data base specialists, has been the "life cycle" of feasibility study, requirements analysis, system design, detailed design, programming, testing, and use. Unfortunately, this approach is often very time-consuming, yet does not produce the system required by the users. When it is "complete," an additional maintenance phase becomes a major activity, as the system is modified to meet user requirements.

In many ways, prototyping, although taking a different approach to system development, must use the same basic tactics. The location and origin of data, as well as their transformation through activities, must be understood. The flow of data and its users at different times must be known. The control issues, such as access rights, audit trails, and review points, must be identified. Prototyping makes good use of user know-how in identifying such issues.

There is a third group of people that will prefer the use of a more business-oriented approach to modelling, to ensure that the system meets the major business requirements and gives priority to these, rather than to less important functions. Typical approaches of this type look at critical success factors, the value chain, and competitive advantage.

In the first of these, effort is focussed on identifying the factors that are essential to meeting business goals. The underlying methodology is based on the experience that, if management focuses on a limited number of critical issues that are essential to business success, their chances of success are higher than if they address less focussed goals. On this basis, if it is decided to focus EDM/EWM only on lead time reduction and product cost reduction, there will be more chance of attaining these goals than if a larger and less defined set of goals is chosen. Value chain theory recognizes that a company's activities are made up of a set of activities, such as sales, logistics, production, and service, each of which adds value to the company's product. By understanding where the company is most effective at adding value and focussing resources on such areas, the company can improve its market position. An attempt to use IT for competitive advantage can also be made. These business-oriented approaches help to ensure that EDM/EWM will be used to meet business objectives and not only to solve data management problems.

One of the most important tasks for the manager of the effort to introduce an EDM/EWM system is to resolve the conflicts between the different aims and different technologies of the developers, users, and managers of EDM/EWM in the overall engineering environment. Many individuals will be unwilling to accept new ways of working and new tools.

7.4 DATA FLOW MODELS

Data flow models relate the various activities of the engineering environment to the use and flow of data. Typically, data flow modelling is carried out as a top-down exercise with as many decompositions or hierarchical levels being introduced as necessary. The system is first described at the top level, often called Level 0. Then each element of the top level is separately described in more detail. The process of decomposition can be carried to very fine levels of detail. Two of the most common techniques used to produce data flow models are IDEF/SADT and Yourdon.

IDEF (ICAM Definition Language) was derived from SADT (Structural Analysis and Design Technique) for the U.S. Air Force's ICAM (Integrated Computer Aided Manufacturing) project. It is hierarchical, initially working with data at a high level, then detailing it at successively lower levels. There are three IDEF models. IDEF0 can be used to describe the functions and associated data flows of the engineering environment. IDEF1 models the structure of data entities that are connected together to show the relationships between entities. A data dictionary is built up describing data entities, their attributes, and their relationships. The data dictionary gives a top-down decomposition of data corresponding to the top-down decomposition of functions. IDEF2 models the functions and flow of data, taking account of time-dependencies.

IDEF0 provides a hierarchical description of engineering activities. A function is represented by a box characterized by the information on four arrows (Figure 7.4). The input arrow shows the input necessary to perform the function, the output arrow shows the product of the function, the control arrow shows the constraints under which the function is performed, and the mechanism arrow shows the means

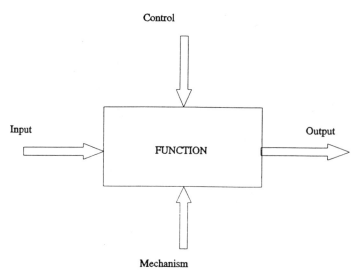

FIGURE 7.4   An IDEF function

by which the function is accomplished. Function boxes are linked on diagrams to show the overall environment. The output of one function is the input for another.

The content of a box (a function) can be decomposed into boxes (sub-functions) at a lower level of diagram. Lower level boxes, of which there are typically very many, show the most detailed information. Development of models can become a very lengthy and difficult task if it is taken down to very detailed levels in the company. IDEF0 typically generates a tremendous amount of data that will not be fully used. Developments such as the IDEF0 triple-diagonal technique make the process more workable and the results more usable. The Yourdon data flow diagram, although different in layout to the IDEF diagram also allows its users to decompose functions into sub-functions, to whatever level is necessary.

The data flow model of the engineering environment describes inputs and outputs of functions, as well as the sequential relationship between functions. However, it does not describe the allowed values and their combinations, nor can it represent time-dependencies, such as synchronization.

7.5 DATA STRUCTURE MODELS

Data structure models describe the static relationships between data. They are generally based on the entity-relationship model introduced in Chapter 6. An entity is any physical or logical object of interest to the engineering organization. Customers, products, suppliers, processes, locations, machines, money, documents, and employees are all examples of entities. An entity does not represent a single item, but a class of similar items, each of which can be characterized in the same way and will be used in the same way in the engineering process. These common properties and characteristics of an entity are referred to as its attributes. Examples of an aircraft's attributes could include color, weight, fuselage length, wing span, and manufacturer. Relationships are associations that describe the link between entities. Graphically, entities are often represented as rectangular boxes, attributes as circles, and relationships as diamonds. Arcs associate entities and relationships to their attributes. Annotations indicate the nature of relationships (one-to-one, one-to-many, many-to-many). Alternatively, single-headed or double-headed arrows can be used.

In the same way that data flow models are decomposed into successively lower hierarchical levels, data structure models also show hierarchy. For example, although a customer's account number is an important attribute of a customer as seen by the billing department, it is not an important attribute for the customer as seen by top management.

7.6 FROM MODEL TO CODE

CASE (Computer Aided Software Engineering) tools are methodologies and software that automate the software development life cycle, reducing development time and improving software quality. CASE tools are used in the analysis, design, programming, and maintenance phases of the software development life cycle. Some of them can also be used to link systems development to business objectives. Most CASE tools provide support for other analysis and design activities, as well as for the modelling activities described above. However, their software for producing data flow diagrams, data structure and entity relationship diagrams, and data dictionaries can generally be used independently of other modules. This type of software is of great benefit in modelling, since the amount of data generated by the modelling activity is all but impossible to handle by purely manual means. However, the ease with which these programs allow modelling to take place should not be allowed to distract engineering users from their prime objective of producing a working EDM/EWM system.

8

The Importance of Data
Exchange and Standards

8.1 DATA EXCHANGE AND STANDARDS

Previous chapters have been focussed primarily on the management of the in-house engineering process and the corresponding data. However, the exchange of data between different functions in a company and between different companies must also be considered. Currently, most companies have several different systems in place and must transfer data between them. In the future, the situation will be the same, as the tendency is not towards one single system that will offer all functions, but towards a range of systems, each of which excels in a particular function. Data exchange is also necessary between systems in different companies. For companies such as car manufacturers, which work with many hundreds of suppliers, it creates a major problem.

Data exchange is a part of the overall engineering process that generates potentially redundant copies of engineering data and therefore requires careful management. It is an activity in which errors are often introduced and time is lost. To reduce errors, costs, and wasted time, as little data as possible should be transferred between systems, or, alternatively, they should make use of the same data base.

Data exchange is not a new topic. Most people in the engineering environment know of the neutral data exchange format IGES (Initial Graphics Exchange Specification), which has been under development since 1978. What is new, though, in the early 1990s, is the increasingly visible effect on data exchange formats of the U.S. DoD's (Department of Defense) CALS (Computer-Aided Acquisition and Logistic Support) program. CALS, initiated in 1985, is organizationally independent of IGES. However, since it addresses the exchange of technical information, it is, in practice, closely related. CALS is primarily oriented towards the military environment, which implies that it will also affect military standards, many of which were developed before computers played such a major role in the engineering environment. Although CALS is oriented primarily to the military environment, it is already having a major effect on civil applications.

The main subjects of this chapter are IGES (and associated acronyms such as VDA and SET), CALS (and associated acronyms such as SGML), and PDES/STEP. Before coming to them, though, three other acronyms—OSI, EDI, and CIM—need to be introduced.

8.2 OSI, EDI, AND CIM

The exchange of information between two entities requires a physical exchange of data and a shared understanding of the informational meaning of the data. The physical exchange of data between two computers from different vendors is addressed by the ISO (International Standards Organization) OSI (Open Systems Interconnection) Reference Model (Figure 8.1). This seven-layer communications model defines a communication infrastructure and corresponding communications services, but does not specify the exact services and protocols to be used in each layer. In principle, all protocols and interfaces for machine-to-machine communication can be defined in relation to the model, thus providing an open framework for networks and networking equipment design and implementation. Examples of standards that comply with OSI are MAP (Manufacturing Automation Protocol) and TOP (Technical Office Protocol). Starting from the lowest level, the OSI levels are named physical, data link, network, transport, session, presentation, and application. The lower levels are mainly involved in getting bits from one machine to another. The presentation layer is concerned with the syntax and semantics of the information being transferred. The application layer handles activities such as file transfer between different machines. FTAM (file transfer access and management) and X.400/electronic mail are at this level.

LAN (local area network) standards address the lowest two levels of OSI. They include Ethernet (IEEE 802.3), MAP (IEEE 802.4), IBM's Token Ring (IEEE

7	Application Layer
6	Presentation Layer
5	Session Layer
4	Transport Layer
3	Network Layer
2	Link Layer
1	Physical Layer

FIGURE 8.1   The ISO OSI Reference model

802.5), and FDDI (fiber distributed data interface). Ethernet is limited to 10 Mbps (megabits per second). FDDI will be capable of 100 Mbps. Companies use WANs (wide area networks) to transfer data between distant sites and with their suppliers and customers. PSDN/X.25 (packet switching data network) and ISDN (integrated services digital network) are WANs. ISDN and X.25 are implementations of OSI's third layer. ISDN will make it possible to transmit data, voice, and image signals simultaneously.

Electronic data interchange (EDI) is generally accepted as referring to the electronic transmission of business data between the computer systems of two companies. The words of the acronym do not specify that only business data (e.g., purchase orders, invoices, shipping documents) can be communicated, but up to 1990, very little engineering data has been transferred this way. The initial developers and users of EDI (from the late 1970s onwards) were interested in transferring business data. For example, the information most commonly transmitted by EDI in the U.S. automotive industry is material releases and advance shipping notices. Business messages, whether they are related to airline reservation, grocery, automotive, chemical, or general merchandise activities, are generally short and numerous, unlike transfers of engineering information, which are generally long and infrequent. As a result, an EDI network architecture designed for communicating business messages would not be efficient for transferring engineering information such as large CAD files.

Electronic data interchange of engineering information will become more frequent during the 1990s, and specialised networks and protocols will be required. Currently, relevant standards include ANSI X12 electronic business data exchange, and ODETTE (Organization for Data Exchange by Teletransmission in Europe), a neutral file format used by European automobile manufacturers. The United Nations Economic Commission for Europe (ECE) has developed the EDIFACT (EDI For Administration, Commerce and Transport) standard to reconcile its own guidelines for Trade Data Interchange (TDI) and the US X12 standard.

CIM (computer integrated manufacturing) will only be feasible for most companies if standards are developed for the computing environment. One of the ESPRIT (European Strategic Program for Research and Development in Information Technologies) activities set up in 1984 was the Computer Integrated Manufacturing Open Systems Architecture (CIM-OSA) project. It recognized the need for international standards for the design of CIM system architectures, as well as the importance of efficient data management techniques. The project's aims included the development of a manufacturing enterprise model, an information reference model, and a CIM implementation reference model. The CIM-OSA framework has three architectural levels, known as generic, partial (industry-specific), and particular (company-specific). It has three modelling levels, known as Requirements Definition, Implementation Description, and Design Specification.

The four views required are function view, information view, resource view, and organization view.

CIM is clearly related to the use of computers and information in the engineering function. Developments in CIM will impact EDM/EWM systems, just as developments in EDM/EWM will affect the overall CIM picture.

8.3 DATA EXCHANGE

The importance of exchanging engineering data is such that significant international effort has been expended in the effort to develop an appropriate standard. There are many hundreds CAD/CAM systems available on the market, as well as many thousands of other engineering programs. Each has its own way of representing and storing data. As a result, when, for example, Company X sends data from its CAD/CAM system A to Company Y, which uses CAD/CAM system B, there is no guarantee that system B will be able to understand the data. There are three solutions to this problem. The first is to use a "direct translator" (Figure 8.2), the second is to use an intermediate neutral format (Figure 8.3). The third, adopted by some large companies, is to try to force suppliers to use exactly the same system. If successful, this latter solution would remove the data conversion problem. However, for small suppliers working with several large companies, this approach can be ruinous as they can neither afford to invest in numerous systems, nor have available specialists skilled in the use of each system.

A direct translator running, for example, in Company Y would be capable of accepting the data in the format of system A and translating it into the format of system B, where it could then be used. Such a translator could be referred to as an "AB translator." To transfer data back to system A, a "BA translator" would be needed. Direct translators have been developed and can be very efficient. Their major disadvantages are their need for modification each time that the data representation of the systems is changed, and their number. Two translators are needed for each pair of systems. For n systems, $n(n-1)$ translators are needed. For 50 systems, 2,450 translators are needed. Although it may seem unlikely that any one company would use so many systems, in 1989 one European car manufacturer and its suppliers together had more than 40 different systems.

International standardization efforts have concentrated on the other approach—the intermediate neutral format. In this approach, each system is equipped with a program known as a preprocessor, to convert data from its own format to the neutral format, and a program known as a postprocessor, to convert data from the neutral format to its own format. In the example of a transfer from system A to system B, the preprocessor of system A translates data from its own format into the neutral format, and the postprocessor of system B translates the neutral formatted data into its own format. For two systems, two preprocessors and two post-processors are needed.

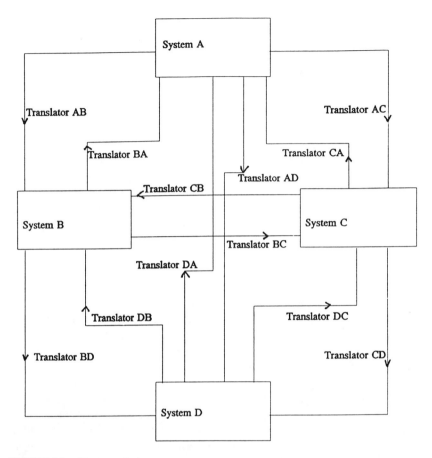

FIGURE 8.2   Direct translation

In the general case of n systems, a total of 2n programs is needed. When n is large, this is significantly smaller than the n(n-1) programs needed for direct translation. Where the neutral format is sufficiently rich in entities, stable over time, and efficient in operation, no doubt the neutral format would be the preferred solution. However, until now this has not always been true for all cases, and a significant number of companies continue to use direct translators.

The neutral format approach raises some questions that are of general interest to those involved in engineering data management (e.g., what should the neutral format include?). Should the neutral format be seen only as a data exchange format, or also as a data storage format? This latter question links directly back to the question of how an EDM/EWM system should store its data.

Before describing the efforts of international standards organizations to define

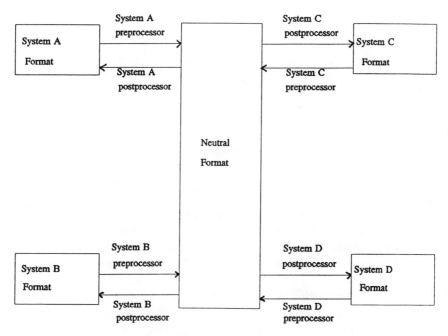

FIGURE 8.3   An intermediate neutral format

a neutral format, it should be noted that some commercial organizations have developed their own "standards." Examples are DXF (Data Exchange Format) from Autodesk, and ISIF (Intergraph Standard Interchange Format) from Intergraph. As these are widely used systems, they can help resolve the data exchange problem for some companies. Another solution that is sufficient in some cases is to use a plot format, such as Hewlett Packard's HPGL. A related subject is the distribution of "part catalogues" on magnetic media or over a network. Ideally, distributors would use a neutral format, but in practice it is interesting to see that they often represent their parts in the native formats of several popular systems.

8.4 FROM IGES 1.0 TO IGES 3.0

IGES was first proposed as an ANSI (American National Standards Institute) standard in 1978. Version 1.0 was adopted in 1981 as ANSI Y14-26M. IGES 1.0 mainly addressed the exchange of mechanical engineering drawings and wireframe three-dimensional models produced by CAD/CAM systems. It was intended to meet the need to communicate data between CAD/CAM systems. It met the need of communicating data between the CAD/CAM systems of the 1970s. In the early 1980s, many potential IGES users found it did not meet their requirements.

In the late 1960s/early 1970s, various other activities that would provide additional specifications for a neutral format were also taking place. CAM-I's (Computer-Aided Manufacturing International) GMP (Geometric Modelling Project) led to XBF (Experimental Boundary File). NASA's (National Aeronautical and Space Administration) IPAD (Integrated Program for Aerospace Vehicle Design) highlighted the importance of engineering data modelling, data exchange, and data management. The U.S. Air Force's ICAM (Integrated Computer-Aided Manufacturing) PDDI (Product Definition Data Interface) project investigated the need to include more than geometric data in the neutral exchange format.

IGES 2, which, as IGES 3.0, was to become the ANSI standard in December 1986, underwent modification and improvement. In this form it is also included in MIL-D-28000.

In Germany, VDA-FS (Verband der Automobilindustrie Flächen Schnittstelle) was developed by the German Car Manufacturers Association as a simple and efficient standard for the transfer of a limited number of entities—in particular, those not addressed by IGES. In 1984 it became the German national standard DIN 66301.

In the French aerospace industry, similar work was carried out, leading in 1985 to the adoption of SET (Standard d'Echange et de Transfert) as French national standard AFNOR Z68300.

8.5 FROM IGES 4.0 TO PDES/STEP 1.0

Back in the United States, development of the following version of IGES (Version 4.0, which was released in 1988) was under way, as were investigations of the next generation of neutral data exchange standards. These had started in 1984 and were to take the name PDES (Product Data Exchange Specification). As a result of the wide international interest in the subject, the ISO set out to develop an internationally accepted standard named STEP (Standard for the Exchange of Product Model Data), under the responsibility of TC184/SC4/WG1 (Technical Committee 184/Steering Committee 4/Working Group 1). STEP will be ISO 10303.

PDES Version 1.0 was to become the working draft for STEP Version 1.0 by 1988. The first draft proposal, which ran to more than 1000 pages, was rejected as incomplete and too voluminous, and a revised version was scheduled for publication in 1991. If approved, this could be available as a Draft International Standard in 1992.

In 1990, the IGES/PDES Organization (IPO) Steering Committee changed the definition of the acronym PDES to Product Data Exchange using STEP. The change reflects IPO's support of the international standard. The original ISO Working Group was reorganized, and its functions divided up between six working groups (WG2 to WG7). The overall standard for STEP was divided into a number of relatively independent parts that may be published separately.

PDES Inc. was set up in 1988 to help advance development and use of the standard. Participating members include major aerospace, automotive, and computer manufacturers.

8.6 EDIF, VHDL, CAD*I, AND CALS

However, before jumping forward to 1990, two other activities that started earlier have to be addressed. One is EDIF (Electronic Design Interchange Format), which was started in 1983, and the other is CALS, which started in 1985. Whereas IGES was mainly developed for mechanical engineering applications, EDIF was developed for the electronics sector. It is intended for the transfer of data, such as logic models, schematics, netlists, and mask layouts. Also in the electronics sector is VHDL (VHSIC Hardware Description Language). VHDL is a DoD-supported standard electronics design language for VHSICs (Very High Speed Integrated Circuits).

In 1984, work started on an ESPRIT project referred to as CAD*I with the objective of defining a CAD data exchange interface, and interfaces to data base systems and other applications. This project made important contributions towards a high level data specification language (HDSL), a neutral file for product geometry, and an interface to finite element analysis applications.

In 1985, the CALS program got under way with the intention of automating and integrating contractor processes for generating technical information and improving DoD capabilities to communicate, use, and store technical information in digital form. By interfacing a contractor's computer-aided design, manufacturing, and support systems with similar DoD systems, major productivity gains could be achieved. In addition to engineering drawings and CAD models, CALS also addresses operational and technical manuals and the logistic support activities associated with maintenance and the procurement of spares.

CALS defines the conditions under which a contractor will be able to exchange information with the DoD and with other contractors. The DoD spends many billions of dollars each year on technical documentation. The volumes involved are high. The DoD has several hundred million engineering drawings in repositories spread around the United States. The technical documentation for modern military aircraft runs to nearly 1 million pages. Their several hundred on-board computers run tens of millions of lines of code. Warships put to sea carrying 25 tons of documents. The weight of the paper documentation describing a helicopter exceeds the weight of the helicopter.

A phased approach to the implementation of CALS was chosen. In Phase 1, islands of automation would be grouped and integrated. In Phase 2, the longer term goal would be of integrated, distributed databases accessed by gateways. A typical initial grouping would lead to a product definition database, a technical support database, and a logistic support database.

The development of a product definition database will link into IGES/PDES/STEP activities. However, as these standards are not yet fully defined, a medium term goal is the electronic transfer of digitized (rasterized) engineering drawings. This activity has its own associated data exchange standards, such as CCITT (Consultative Committee for International Telegraphy and Telephony) Group 4 raster scan standards. The first phase of CALS also calls for use of the CGM (Computer Graphics Metafile) standard for vector drawings, and IGES for engineering drawings. Unlike IGES, CGM, a relatively basic 2-D representation, is concerned with the description of the image, not the product. The image is described in a device-independent way through graphics primitives (such as lines, texts, and fill areas) that can be further defined through attributes (such as the color or style of a line). CGM is the subject of ISO 8632.

Other standards in the engineering environment include SGML (Standard Generalized Mark-up Language) and ODA/ODIF (Office Document Architecture/Office Document Interchange Format), which address compound documents containing graphics and text. SGML is particularly important for electronic publishing. It provides a methodology for separating out the structure, content, and format of a document. Once elements of a document, such as paragraphs, titles, and headings, have been marked up, they can later be selectively retrieved and published in a variety of styles. ISO 8879 corresponds to SGML. The Office Document Architecture standard is ISO 8613. Part 5 deals with document interchange formats. One of these is ODIF. Another is ODL (Office Document Language) and SDIF (SGML Document Interchange Format)—parts 6, 7, and 8 of ISO 8613 address character, raster graphic, and vector graphic content respectively.

The success of CALS depends on standards, and it has been supported by the NBS (National Bureau of Standards), now the NIST (National Institute of Standards and Technology). As well as providing input to PDES/STEP activities, CALS is also affecting standards in the DoD environment. MIL-STD-1840A Automated Interchange of Technical Information outlines these. MIL-D-28000 addresses IGES guidelines. MIL-M-28001 addresses SGML issues, text mark-up requirements, and specifications for technical manuals produced by computer-aided publishing systems. MIL-R-28002 addresses raster images. MIL-D-28003 addresses CGM.

Future CALS activities should include the on-line access of distributed databases. MIL-STD-1388-2B addresses record management from file processing to relational data base management. MIL-STD-1388-3B will support the longer term goal of neutral access to distributed databases through a tool such as Structured Query Language (SQL), the subject of ISO 9075.

MIL-T-31000 is a new specification for the technical data packages that support the acquisition strategy, engineering, production, and logistics support of a product. The technical data package includes product definition items, such as drawings, specifications, performance requirements, lists, operating instructions, and software, and product management items, such as source control drawing approval requests, drawing number assignments, process descriptions, and quality plans.

8.7 CCITT GROUP 4 AND TIFF

Before returning to PDES/STEP, the above-mentioned standard CCITT Group 4 and another raster format standard, TIFF (Tagged Image Format Files) will be described. The TIFF specification describes storage of CCITT Group 4 compressed data.

Typically, engineering drawings are scanned at about 200 dpi (dots per inch). Each "dot" or "pixel" is stored in a bit. A white dot could be stored as "zero," and a black dot as "one." If all pixels are stored, an E(A0) size drawing takes up about 8 MB. This is expensive and probably unnecessary, since techniques exist to "compress" the data. For example, a run of white dots can be replaced by a number corresponding to the number of dots. A technique called Modified Huffman coding achieves further compression by using codes to signal commonly occurring runs. The Modified Read technique, instead of storing an individual scanned line, may only store the difference between the line and the previously scanned line. The information in the 8 MB can probably be "compressed" with such techniques to about 1 MB. CCITT T.6 (Group 4) is the standard adopted under the CALS program. It uses a combination of the Modified Huffman and Modified Read techniques, and gives typical compression factors between 10 and 50. (The standard for fax transmission is Group 3.) Special compression techniques are needed to handle large documents, such as engineering drawings. "Tiling" is such a technique. It breaks a large document up into smaller pieces that are processed individually and then reassembled.

Just as data exchange and communications can benefit from standardization, so can the user interface for applications using engineering data. The X Window System is becoming the standard WIMP (Window, Icon, Menu, Pointer) graphic user interface (GUI) for stand-alone and networked environments. It has its roots in a 1984 MIT (Massachussetts Institute of Technology) project. Since 1988, the X consortium has been responsible for the X standard. X11 is the X window protocol.

PHIGS (Programmer's Hierarchical Interactive Graphics System) is another proposed standard that will affect the engineering environment. It provides a common programming base for graphics programming so that applications can be used on otherwise incompatible graphics devices.

8.8 PDES/STEP

Coming back to product definition standards, IGES has continued to evolve with Version 4.0 in 1988 and Version 5.0 in September 1990 (NISTIR 4412). At this point, before reintroducing PDES, it is useful to look in more detail at IGES. IGES was intended to address graphics and geometry, and not complete product definition. Versions 4.0 and 5.0 of IGES cover entities ranging from simple geometric and dimensioning entities to three-dimensional surface and solid geometry entities. IGES uses an 80-column card format. Each entity has a defined code (e.g., 110 for a line, 116 for a point, 114 for a spline surface, and 128 for a B-spline surface) and a defined format for information describing the entity (e.g., three points for a circle).

The intention of PDES is to be able to transfer all product definition data (not just the geometry) into a logical, computer-readable format. PDES has a three-layer structure. The lowest layer—the physical layer—contains the detailed data on individual files. The middle layer is a conceptual schema describing the entities, rules, and relationships that make up the engineering environment. At the top layer—the application layer—are the sub-schema and data of the application programs that need to communicate data.

This approach, similar to the conceptual schema approach to data bases described in Chapter 6, frees the application from the details of the physical implementation of data.

The conceptual, or logical, layer lies at the heart of PDES, and the following list of some of the types of information it contains shows that it is much more complete than IGES:

Types and functions
Miscellaneous resources
Geometry
Topology
Shape representation
Features
Shape representation interface
Tolerances
Materials
Presentation
Product life cycle
Applications
Product manifestation
Product structure
Electrical applications
Architecture/engineering/construction applications
Ship applications

Analysis applications
Data transfer applications

There are second level entries for many of the entries in the above list. For example, the second-level entries for the "Presentation" entry are:

Introduction to presentation
Basic presentation entities
Libraries
Tables
Table of line styles
Table of surface styles
Instances
Composite structures
Components of View with annotation

The "Geometry" application area includes geometrical entities such as curves, surfaces, CSG (constructive solid geometry), and B-rep (boundary representation) solids.

It is worth noting the presence of "Product Life Cycle" and "Product Structure" within the above list, as these will relate directly to project management and product configuration.

Underlining the importance of the data modelling techniques described in Chapter 6, PDES/STEP makes use of formal data modelling techniques. The conceptual schema is mainly modelled in the EXPRESS data description language, a Pascal-like declarative language that evolved from the PDDI project. The model represents the formal definition of the entities that make up PDES/STEP. IDEF1X, which evolved from the ICAM program, was at one time the preferred language for modelling the conceptual schema. NIAM (Nijssen Information Analysis Method) has also been widely used.

The PDES/STEP physical file structure is a sequential ASCII character-based file. In principle, it should be possible to map automatically from the conceptual schema, through the EXPRESS definition, to the physical exchange file.

Four implementation levels have been proposed for the PDES physical layer:

Level 1, a passive file exchange, is similar to the IGES implementation. Data to be exchanged from one application to another is preprocessed, transferred in file format, and then post-processed.
Level 2 is similar to Level 1 in that it addresses data exchange and not data sharing. However, in Level 2, the "exchange file" is memory-resident. To exchange data, the applications transfer data to and from this memory-resident file. The applications still have their own versions of data in their own native format.

In Level 3, the data is stored in a common data base, which the applications access through data base procedures. There is only one copy of the data in Level 3.

Level 4 will be based on knowledge-based techniques and would probably use an object-oriented data base approach. For both Level 3 and Level 4, the intention is to share, rather than exchange, data.

9

Planning to Use EDM/EWM for Competitive Advantage

9.1 INTRODUCTION

This chapter and the following chapter look at the various steps (Exhibit 9.1) along the road towards successful implementation of EDM/EWM. Chronologically, the activities should occur roughly in the order in which they are described. However, many of the activities overlap, and there may well be occasions when the findings of a given activity or the inability to agree on a proposed solution will lead to previous activities being revisited.

9.2 PROJECT START-UP

The successful introduction of EDM/EWM is a long-term, cross-functional process. It will almost certainly be hampered by a general lack of understanding and knowledge of EDM/EWM, the usual resistance to change, departmental interests, different functional priorities, cost justification issues, and various on-going projects related in some way to engineering data and/or workflow. To help overcome these retarding forces, the introduction of EDM/EWM needs the full support of top management throughout many years. Only top management will be able to resolve the interdepartmental issues that will arise from company-wide use of EDM/EWM. Top management involvement is necessary for long-term, costly, and strategic projects such as EDM/EWM. However, top management cannot be expected to be involved in all the details of selecting and implementing an EDM/EWM solution. This task should be delegated.

Top management should appoint a task force (or engineering information and workflow study group) that will be responsible for developing an EDM/EWM

PHASE 1

        Project start-up

        Understanding EDM/EWM

        Description of the current situation

        Competitor's use of EDM/EWM

        User requirements

        Business objectives and EDM/EWM

        Understanding EDM/EWM solutions

        The IT competence of the company

        Sizing and grouping

        The features required

        Scenarios for EDM/EWM

        Prototyping and benchmarking

        Developing the EDM/EWM strategy

        Organizational issues

        Architectures

        The overall implementation plan

PHASE 2

        Detailed design and implementation

PHASE 3

        Use of EDM/EWM

EXHIBIT 9.1   Three phases for EDM/EWM implementation

strategy and implementation plan. The EDM/EWM strategy will provide a high-level definition of the role of EDM/EWM in the company, and outline the corresponding organizational and policy requirements.

The task force should include competent and powerful individuals from all the functions that will create, transmit, or use engineering data. These functions will include R&D, design engineering, manufacturing engineering, marketing, sales, F&A, manufacturing planning, manufacturing, logistics, maintenance, and IT. It will also be useful to include in the task force an individual who will represent the interests of suppliers and customers and any other users of engineering information outside the company.

The task force should include not only representatives of users of engineering information, but also computer systems specialists. Although the IT (ex-EDP) function has not, in many companies, been very much involved in engineering computing, it can provide important support to the task force. IT specialists may know little about engineering information systems, but, on the other hand, few engineers know much about the latest developments in IT, and are unlikely to find an appropriate solution unaided. As users and IT specialists often have difficulty in working together, it is important that only individuals who can work well together are selected for the Task Force.

Top management must define the objectives and authority of the task force. The objective of the task force should not be to select a particular EDM/EWM solution, but to make sure that engineering information is used and communicated efficiently and effectively, with the aim of improving the company's overall business results. The task force will typically be led by a top manager, such as the engineering VP, or someone of similar stature.

Implementation of EDM/EWM is a long-term activity that should take place only once. It should be run as a well defined, company-wide project. A task force secretary, reporting to the task force leader, should run its day-to-day activities. The Task Force secretary may work full-time on the EDM/EWM project, whereas the Task Force leader will only have part-time involvement. A project plan should be drawn up showing the major milestones, resources, activities, and costs. The plan should be kept up to date and visible. Top management, in particular, should know what progress is being made, and be kept aware of any problems that may arise.

An early activity of the task force is to carry out a feasibility study to demonstrate potential scope, areas of application, costs, and benefits. Another is to inform users of engineering information of its activities. If users are aware of potential changes before they take place and are asked for their help, they are usually willing to contribute and, by doing so, become involved and identified with the eventual solution.

Implementation will be a three-phase process. The first stage, in which the task force plays a major role, leads to the development of an EDM/EWM strategy that addresses technological, organizational, and financial issues. It also provides a long-term implementation plan. The second phase starts from the high-level results of the first phase. A rebuilt task force, probably containing some members of the original task force and some other engineering managers and users, will translate the high-level plans into detailed actions. For example, detailed data base and workflow design will be carried out in this phase. The third phase sees the EDM/EWM solution in everyday use. This will, of course, lead to requests for modifications and further improvements.

9.3 UNDERSTANDING EDM/EWM

EDM/EWM is a relatively new subject, and few, if any, of the individuals in the task force will know much about it. Although both the members of the task force and other members of the company will learn much from their participation in the various activities, it will be helpful if they can receive some basic information right at the beginning of the project from an acknowledged specialist in EDM/EWM. Otherwise, it is only too likely that they will go round in circles for a considerable time, with each member of the task force becoming more and more closely attached to the idea that EDM/EWM mainly exists to solve his or her everyday problems. Some will see EDM/EWM as only being a solution to CAD/CAM data management problems. Some will see it as a bill of materials project. Some will see it as being the answer to configuration management and traceability problems. Some will see it as a way of making sure that their favorite procedures are implemented. Others will see it as an answer to connecting PCs to the corporate mainframe. As time goes by, each one will become more and more convinced that they alone are right. The intervention of a neutral, experienced expert at an early stage can prevent this negative and resource-wasting state of affairs from arising.

It may be possible for the task force members to attend an introductory EDM/EWM course so that they can understand some of the basic points of EDM/EWM, and learn together some of the vocabulary specific to EDM/EWM. Useful information on EDM/EWM can also be gathered from books and journals, from conferences and seminars, from demonstrations by vendors of EDM/EWM solutions, and from visits to other companies using EDM/EWM.

Engineering journals generally provide some information about the EDM/EWM market, and from this it should be possible to learn about the typical cost of EDM/EWM implementation and of individual systems. The initial understanding of EDM/EWM should not be restricted to the task force. Management should be kept up to date, as should potential EDM/EWM users. The more that management knows about EDM/EWM, the more supportive it will be of the task force. The more the users know about task force progress, the more supportive they will be, and the less likely they will be to start competing activities and overlapping projects.

The four major areas that the task force needs to understand before proposing a strategy for EDM/EWM are the current engineering process and information environment, the company's business objectives, the users' requirements, and the functionality of the systems available. Study groups may be set up within the task force to address particular issues such as understanding the flow of the current product development and manufacturing process, describing the systems currently in place, and identifying the EDM/EWM activities of competitors. At a later stage, individual study groups could take on tasks such as describing the required systems and estimating the business benefits of EDM/EWM.

9.4 DESCRIPTION OF THE CURRENT SITUATION

The description of the current situation and the understanding of user's requirements, although presented separately here, are complementary activities. The description of the current situation addresses seven subjects:

- The functional use and flow of engineering information,
- The activities that create and make use of the engineering information,
- The organization of engineering,
- The users of engineering information,
- The life cycles of products and projects,
- The management of engineering information and the engineering process, and
- The current methods (both manual and automated) of creating, communicating, and storing engineering information.

Information models will need to be developed to help describe the current situation. In Chapter 7, the need for information modelling was introduced. It is at this stage of the task force's activities that such models will start to be useful. However, since many months or even years can be spent in developing information models, this activity should be kept under close task force control. Although modelling is a useful activity, it can become very expensive and time consuming if pursued to a too-detailed level. Initially, the task force may only need high-level information models, with further refinement only taking place once a high-level EDM/EWM strategy has been defined and it is possible to see where modelling would be most effective. The task force must define the objectives of modelling and the expected end-product, before initiating modelling activities. Where appropriate, existing models can be used.

The task force needs to understand how engineering information is created, used, and communicated within the company. The design engineering department is clearly a major source of information, but there are other sources and many other users, including manufacturing engineers, manufacturing planners, suppliers, customers, and after-sales service staff. The task force can take two complementary approaches to understanding engineering information flow. These are a top-down functional decomposition approach and a "sideways" product path approach that follows product information through successive activities. A particular product may start in the marketing function and then go through conceptual design, engineering design and analysis, testing, detailed design, manufacturing engineering, process planning, tooling, NC programming, production planning, purchasing, sales order processing, machining, assembly, testing, packaging, distribution, and maintenance. Other new products will follow a different route, as will minor changes to existing products. In some cases, the path will lead out of the company to suppliers and partners. In most cases, it will eventually lead back to customers. Detailed workflow diagrams will need to be produced to help understand the workflow and communicate this understanding.

As well as understanding the flow and use of engineering information, the task force should also address the individual activities that create or use this information. These "application" areas, such as engineering design, process planning, and NC programming, will probably be partly automated, but still have a significant manual content. The task force needs to understand the information needs (input, processing, output, and storage) of each application area. These applications, which are part of the overall engineering information process, will need to be integrated into the overall EDM/EWM solution. The different structures of engineering information such as bills of materials, assemblies, and parts lists should be identified, and other associations such as product/drawing relationships clarified. The task force will begin to understand the way that "packets" of information are created, modified, and moved between activities.

As the individual activities are examined, the task force will begin to understand not only the information needs of each activity, but also some of the parameters concerning the volume of information involved. These will include the number of existing products, parts and tools, the annual number of new products, parts and tools, the number of software versions, the number of modifications, the number of new and modified drawings and other documents, the number of drawings released daily, typical design times, the number of engineering changes, the timing of engineering changes, the time taken to process engineering changes, and the number of levels and constituents of bills of materials. The amount of data created each year by computer-based systems will be found and analyzed. The volumes and lifetimes of products need to be understood. It is important to know these parameters, as the EDM/EWM requirements of a company oriented towards low-volume, short-life, fast-changing products will be very different from those of a company oriented towards mass-produced, long-life, stable products.

The current organization of the company needs to be described from the point of view of engineering information. This will show the number of users and their location, both geographically and functionally, and the way they store and communicate information. It will show where data is stored and how it is shared.

The sources and users of engineering information should be identified. This exercise will highlight the large number of users who are not within the conventional definition of the engineering department. An attempt should be made to understand how users create, access, modify, store, and communicate information. Some surveys indicate that drafters spend more time looking for information than drafting, and that designers spend more than one-third of their time looking for information. Many users spend most of their time moving, preparing, and controlling information, and little time adding value to it. If users neither add value to information, nor refer to it, perhaps they do not need to receive it. The access needs and rights of users and groups of users need to be understood by the task force. Shared and redundant data needs to be identified. Data standards and data ownership have to be understood.

As well as looking at information use and flow along product paths, the task force should also look at the structure of product paths, and identify the typical milestones, events, and management activities along these paths. There will be different structures for different product lines, and, no doubt, some individual projects will follow specific rules. This information will help to provide the generic product life cycles, which are a key component of effective EDM/EWM. The project management techniques in use should be identified. Project status and review needs should be identified. This information will be needed for workflow management.

The management of engineering information, in particular at departmental boundaries, needs to be understood. Data security and data integrity issues will need to be addressed. Existing data management systems need to be examined. The transition rules between the different states of information must be classified. The rules vary along the product life, from the initial product concept, during which the information's owner can modify it at will, to the time when the product is in the customer's hands, and information can only be modified if strict conditions are met.

Review, release and change processes need to be understood. The task force should discover how many engineering changes are made, as well as the way they are made and recorded. The time and effort required to carry through changes should be brought to light. The roles and rights of users and managers at change and release time must be understood.

The current methods of creating, communicating, and storing engineering information need to be understood and quantified. The cost structure for preparation and distribution of documents will need to be understood. The volumes of drawings and other documents in storage and under modification are important parameters. The quantity of information communicated will be an important parameter for network design. There may well be comparatively little known users and repositories of engineering information that would be ignored without a wide-ranging analysis.

Particular attention should be paid to areas outside the traditional engineering area. There will probably be engineering information in the production planning systems (such as MRP 2), NC part programming systems, process planning systems, analysis programs, test systems, quality control systems, office automation systems, F&A systems, and spreadsheets. Suppliers, partners, and customers may also store and use the company's engineering information.

An inventory of existing IT systems related to engineering information should be drawn up. As well as obvious candidates, such as CAD and CAM systems, the list should include systems used in analysis, CASE, project management, technical publications, documentation management, configuration management, and release control. The corresponding computers should be identified. The programs in use should be listed and analyzed. Their use of engineering information needs to be understood. Any data base management systems in use should be closely examined. Information transfer between systems should be described. Transfer of information to and from supplier and customer systems should be included in the survey. Use of EDI should be noted, along with its data characteristics. A list of current engineering, CIM, and IT projects that may affect engineering workflow and engineering information should be drawn up. The priority of some of these may have to be modified. It may even be that some of them will be abandoned.

As the task force learns about the current situation, it may well identify activities that are illogical and/or wasteful. Perhaps some administrative procedures will be found to be unnecessary or duplicated. Some documents or drawings may not actually be used. The task force should signal such findings to management. It may be possible to initiate short-term actions to correct them. However, it is not the role of the task force to carry out such actions.

9.5 COMPETITOR'S USE OF EDM/EWM

To complement their understanding of the way in which the company uses and manages engineering information, the task force should try to evaluate the way that competitors are organized, and the steps that they have taken towards implementing EDM/EWM solutions. The task force should also look at the way that competitor's suppliers and customers are using EDM/EWM. This may help the task force to understand competitor's behavior, and may provide opportunities to create new relationships. In some cases, competitors will be willing to invite the task force for an open exchange of information. If this is not possible, something may be learned from presentations made at conferences and seminars, or from articles in journals and magazines. Some industry associations, recognizing that EDM/EWM will be vitally important to their members, may run seminars in which experience is exchanged among members, and external experts give information on EDM/EWM developments in other industries.

9.6 USER REQUIREMENTS

Another major source of information for the task force will be the current and potential users of engineering information. However, many of these will probably never have thought about the overall EDM/EWM requirements of the company, so the task force will have to develop specific techniques, to find the necessary information. Task force study groups may be set up to interview individual users and groups of users. Some of these will be from individual functions, and some from cross-functional organizational structures, such as product teams. In general, users are the best source of information about everyday activities, the information that is used in such activities, and short-term improvements. Users often find it more difficult to address longer-term and cross-functional issues. The task force should encourage users to discuss their problems with data access, use, and management, since it is often the users who are the people closest to the problems of inaccurate, erroneous, and late data.

In some cases, the task force will use a suitably-designed questionnaire to gather specific results from users. In other cases, this may not be necessary, but an interview document will need to be produced to encourage individual interviewers to remember to ask the most important questions. Sometimes, a top-down hierarchical breakdown of information requirements may be a suitable modelling approach. The models produced can be easily understood by most users. The amount of detail in the model can be progressively increased until a user feels that all information use and flow is shown on the diagram. Other users can be asked to comment on the model, and models produced by users in neighboring activities can be put together, to see how information is transferred.

Another approach is to use the entity-relationship model. Again, individual users can be helped to produce their "picture" of the information they use. Once the entities have been identified, the next step will be to identify their attributes. This approach, like all modelling activities, can be very time-consuming if carried out to the finest level of detail. The task force should decide how much time can be spent on modelling, and then define the most important activities to be modelled and the amount of detail that is required. During the early stages, the task force may not need very detailed information. The models developed at this stage can be kept, and then worked on in more detail in phase 2 of the EDM/EWM implementation process, when a solution has been chosen and detailed logical data base design takes place.

Even during the earliest stages, it will be helpful to use computer-based tools to manage the information models. It may also be possible to develop prototypes that users can run in everyday work, thus giving them a better feel for the way that EDM/EWM can assist them, and helping them perhaps to identify necessary modifications and improvements.

The users should also be encouraged to explain their view of the overall process. The task force will learn how they receive work, how they know when to start working on a project or to change to another project, the organizational hierarchy, the release procedure, and so on. They will see where information is created, how it flows between users, the way it is used, distributed and stored, and the corresponding management actions.

The task force must not restrict itself to understanding the requirements of users in the engineering department. The needs of other users, including those in the marketing function and on the shop floor, also have to be addressed. Similarly, the needs of departmental managers are as important to understand as those of users, such as part programmers.

The task force may detect user concerns about the implementation of EDM/EWM. These can range from fears of loss of status, job satisfaction, and job, to worries about retraining, reassignment, and regrading. If such attitudes are prevalent, the task force will need to increase user involvement and improve communications with users.

9.7 BUSINESS OBJECTIVES AND EDM/EWM

An extremely important input to the task force that can only be obtained from top management is information on the business objectives of the company as they relate to EDM/EWM. In most cases, top management will not be able to give this answer directly. The task force will be able to piece it together from information that is more readily available, such as the company strategy, the corporate critical success factors, and the individual IT, R&D/engineering, and manufacturing strategies. The relationships between business objectives and EDM/EWM may also become apparent from the issues raised and the concerns expressed, by top management, when discussing EDM/EWM and related subjects. The task force should try to identify the four or five factors that seem to be most important for management. These could include, for example, the need to be CALS-compatible for work with the DoD. There could be a need to reduce lead times significantly or to improve product quality and reliability. There could be specific problems that have to be avoided, or relationships with powerful customers that need to be improved. There may be the intention to delete some product lines or to develop new, or improved products. There could be plans to change the way that clients and markets are addressed. Product development costs and product costs may be too high. Support staff costs may need to be reduced.

The factors may be very closely linked to business strategy. A company intent on market dominance through lowest unit cost may have very different EDM/EWM requirements from one that will provide highest performance products to specification in a speciality niche market. Performance-driven companies emphasize continually evolving designs, whereas inflexible, lowest-cost manufacturers will want to freeze product designs.

These top-level issues are just as important for EDM/EWM as the "low-level" user requirements. If, for example, a particular product or process is to be abandoned, it may be excluded from the analysis. If the company has decided only to purchase computers from three known vendors, the task force should not waste time developing solutions from other vendors. If the company has an "Open Systems" approach, the task force may be able to look at a wider range of suppliers.

If possible, the information obtained from management should be quantified. It is not enough to know that profitability and market share must be increased. Quantified information has more meaning, and can be used both as a target and as a measure of progress. Management may want to focus use of EDM/EWM on reducing product cost. Management may want to introduce major business programs, such as total quality management, just-in-time, or design for manufacture. Management may be aiming to reduce the lead time in the engineering department by 50 percent. If the task force believes that EDM/EWM cannot help achieve this goal, it should inform top management why they see this goal as unrealistic. In such a case, the task force might be able to propose another solution, perhaps involving the removal of a bottleneck in the process.

The business objectives provide a clear business focus for the task force. This will help them greatly, and, in particular, it should prevent them from drowning in the sea of information that they will produce. Without the business objectives, the task force can all too easily produce a "technological solution" that is of no benefit to the business. With the business objectives, the task force has clear targets in site, and can focus its activities and prioritize its recommendations. Knowledge of the business objectives will help the task force to balance the perhaps divergent needs of improved management of engineering drawings, CAD/CAM data, alphanumeric engineering documents, alphanumeric/graphic documents, and engineering workflow.

Some top managers may believe that an EDM/EWM system is going to solve all the problems associated with their engineering activities. It should be pointed out to them very clearly, while discussing business objectives, that an EDM/EWM system can only support their drive to improve business performance. If other areas (Exhibit 9.2) are not addressed, the EDM/EWM system is unlikely to provide the required business benefits, and its major effect will be to increase costs.

9.8 UNDERSTANDING EDM/EWM SOLUTIONS

There are many solutions offered to help improve the management of engineering information and the engineering process. The task force may eventually decide that there is no one vendor capable of meeting all of the requirements of the company, and a multi-vendor integrated solution may be necessary. Some systems are primarily oriented towards management of CAD/CAM data. Others focus on the management of engineering drawings or on the management of alphanumeric engineering documents. Some systems are oriented towards the management of metadata, and others to the underlying data. The task force will need to examine

Top Management

Engineering Strategy

Corporate Culture

Engineering Organization

Processes

Products

Human Resources

Customers

Suppliers

Technology

Innovation

Engineering Information

Engineering Systems

EXHIBIT 9.2   Target areas for engineering performance improvement

the hardware, software, and communications aspects of proposed solutions. The basics of data base theory and data modelling will need to be understood. Although some information on these subjects may be picked up from seminars, books, and journals, often the best way to learn is from direct contact with vendors and existing users, or by implementing a pilot solution.

Some systems require little customization; others require much more. A micro-computer-based EDM/EWM system for half a dozen engineers in a small company may be purchased for less than $10,000. In a large company, with several hundred potential EDM/EWM system users, the initial EDM/EWM system cost may exceed $100,000, and, with user licences for additional users costing several thousand dollars, total system purchase cost may exceed $1,000,000. It is import-ant to understand the total cost of a system over its expected lifetime, taking account of the cost of initial customization, the cost of customizing upgrades, and maintenance costs. Corresponding organizational costs, such as reorganization, training, and the development of new procedures should also be taken into account.

9.9 THE IT COMPETENCE OF THE COMPANY

The task force should quantify the information technology ability and resources of the company. The aim of this activity is to avoid proposing a solution that requires

IT support that the company's IT function is incapable of supplying or managing. The company's IT function may be centralized, or there may be a "central" IT group reporting to the F&A department, with local IT support teams in each individual function. In many cases it will be found that the IT group sees F&A as its primary client, and all other departments as secondary clients, yet is unable even to provide F&A with the right service. Similarly, IT support teams in individual functions are often unable to address strategic issues, as they are overwhelmed with mundane tasks.

In such an environment, it may be difficult to see where the resources for major EDM/EWM development and implementation work will come from. Top management must decide whether new resources should be hired, or existing staff assigned new roles. If neither of these are possible, the task force may be led to solutions that, once implemented, will require very little IT experience and support for on-going use.

9.10 SIZING AND GROUPING

The task force will collect a mass of data about the engineering information requirements of the company. It must not present all of this to top management. Instead, the task force must distill it into a form in which management can easily understand it and recognize the way in which it relates to the company and to the objectives set for the EDM/EWM project. The task force can try to produce a one-page overview that contains all of the most important data. This should show how the business objectives are to be met. It could include the major functions and systems currently involved, with an indication of the volume and type of information created, used, and communicated. It may well be that such a figure will be unbalanced and that a more logical grouping of activities and information can be found. Starting from this one-page, top-level picture, the task force can then develop one-page pictures of each of the major activity and information groups. These will show some of the lower-level activities, the systems involved, the users, and the volumes and frequencies of data creation, use, and communication.

A hierarchical approach should also be taken to sizing. At the top of the hierarchy are the business objectives and the customers. At the next level down are the volumes of existing, new, and modified products. At the following level down are parts and subassemblies. Some of these may be in constant use, and others only referred to occasionally. Similarly, the number of products is important high-level information. The number of documents, files, and drawings in production and use is directly linked to the number of products and projects, but is of less interest to top management and can be changed by modifying the engineering process.

9.11 THE FEATURES REQUIRED

The task force needs to develop a list of features believed necessary for an EDM/EWM solution. Exhibit 9.3 shows some typical features. Preparing such a list helps to provoke discussion and get agreement on the scope of the solution to be proposed. It can also show that the various issues raised have at least been understood and noted, even if they may not be answered in the eventual solution. The task force should indicate if each individual feature is an essential part of the solution, and, if not, assign it a relative importance factor. The first list that the task force produces will probably have very many entries. These will have to be examined closely, grouped, and checked, to ensure that duplicate or conflicting entries are not included.

9.12 SCENARIOS FOR EDM

Once the task force has completed the above steps, it should be in a position to identify potential solutions, or scenarios. These should take into account not only EDM/EWM systems, but also the general environment surrounding engineering information and the processes that make use of it. It will be useful to investigate between three and six scenarios. The task force needs to describe each scenario in detail, along with its strengths and weaknesses. This exercise is extremely useful in gaining an in-depth understanding of a proposed solution. Often, it is by trying to understand the strengths of a scenario that the weaknesses of other scenarios become more apparent. Apart from the purely technical aspects of the scenarios, the task force should also consider financial and organizational aspects. The costs examined should include not only those for purchase, but also those for maintenance, installation, new versions, expansion of the system, and interfaces to other systems. The related benefits to the company, such as decreased product costs and increased sales, should be evaluated. Vendor criteria, such as commitment to EDM/EWM, development plan, ability to develop and upgrade, maintenance record, growth record, user group, delivery time, and availability of technical assistance should be considered.

Among the organizational aspects that will need to be examined and costed are implementation planning, system installation, training, development of procedures, use of standards, standardization, system management and support, security, work methods, work flow, information management, management roles and responsibilities, and potential reassignment of roles and responsibilities.

The task force needs to understand and to be able to communicate the major differences between the scenarios from the business point of view. This is the high-level message it will have to give to top management.

Reduce scrap and rework.

Reduce lead times. Reduce product development costs.

Provide status information on engineering changes.

Provide correct version of information.

Save corporate know-how.

Have potential for use of knowledge-based techniques.

Provide traceability.

Classify and catalogue products and parts.

Manage configurations.

Provide navigation mechanism.

Provide single controlled source for all engineering information.

Search for data on part number, drawing number, release date, drafter.

Handle both CAD and manual drawings.

Provide release procedures. Streamline the sign-off process.

Automatic back-up and archiving.

Data management and control. File management.

Periodic and on-demand reporting.

Reduce volume of paper in circulation.

Relate specifications to design data.

Provide overview of status of engineering activities.

Digitize engineering drawings.

Provide for multiple data definition.

Manage engineering changes and multiple versions.

Control access. Provide instant access.

Provide security, security tracking, and audit trail.

Maintain data integrity. Reduce transcription errors.

Usable by several departments. Have shop floor interface.

Distribute documents electronically.

Link systems together. Link data bases together.

Exchange data with other company data bases and programs.

Handle online user queries.

Engineering information vault. Handle very large volumes of data.

Support standardization.

Be available for maintenance engineers.

Issue documents on optical disk.

Provide simultaneous access to business and engineering data.

Communicate order status to sales staff.

Incorporate good design principles.

Improve communications between engineering and purchasing.

Provide timely, high-quality management information.

Permit controlled partial release.

Provide better cost estimates and bids.

Improve project coordination.

Relate designs to current manufacturing processes.

Attract high-quality personnel.

EXHIBIT 9.3   Typical features required for the EDM/EWM solution

9.13 PROTOTYPING AND BENCHMARKING

As part of their activities in learning about each scenario, the task force may want to gain some hands-on experience with individual systems. Testing a system for a few days, usually at a vendor site, is referred to as benchmarking. It is appropriate for some of the systems that make up the EDM/EWM environment. However, it may be difficult to simulate, in a benchmark, the real-life environment of multiple users, multiple systems, and very large volumes of data. If the task force intends to benchmark competitive systems, the same test should be used for each system. The test should be typical of the work carried out by the company, since the benchmark is not carried out to provide a general evaluation of the system, but to test it for use in the company. The benchmark should be well prepared in advance of the test, and should relate directly to the features that are of most importance for the task force. For complex systems, benchmarking is a difficult activity, and the results should be treated with caution. Poor benchmark performance may result from poor benchmark specification or a failure by vendor personnel to understand the task force's request. A benchmark of a system that is going to play a major role in managing engineering information is extremely difficult and time-consuming in view of the complexity, high volumes, and multiple relationships involved.

To overcome some of the shortcomings of benchmarking, the task force may decide that it is better to build a prototype solution in-house. This will probably involve some payment to the vendor, but will allow the task force to see how the system works when it is in-house and used by company personnel. Pilot studies help to both better understand and to validate proposals. It is often more effective and ultimately quicker to carry out pilot studies than to try to introduce major changes in one fell swoop.

9.14 DEFINING THE EDM/EWM STRATEGY

The task force must develop an EDM/EWM strategy that meets management objectives, and define an EDM/EWM implementation plan. The first step is to develop and describe the strategy and to get top management agreement. In different companies, this process will take place in different ways. In same cases, top management will want to examine the scenarios in detail and then take the decision. In other cases, the task force will be free to define a strategy that top management will then review. In some companies, top management may insist that benchmarking and/or prototyping do not take place before the strategy is agreed. In others, top management may feel that it is useful to have some practical experience of EDM/EWM before finalizing the strategy. Definition of the strategy is all-important, as the rest of the implementation is guided by the strategy. If the strategy is wrong, the implementation will probably fail to meet the objectives.

Before the strategy can be finalized, and before the detailed implementation plan can be developed, the task force needs to clarify three major areas of concern. These are business benefits, organizational issues, and EDM/EWM architectures.

9.15 BUSINESS BENEFITS OF EDM/EWM

The primary aim of EDM/EWM is to improve the company's business position. The task force will know, from its discussions with top management, the particular areas on which it should be focussing. Typically, these will include reduced lead times and reduced product costs. They may also include reduced scrap or reduced rework. Sometimes, management may be looking to increase the amount of time that users spend on productive work, and reduce the time they spend on tasks that do not add value, such as searching for lost drawings and carrying out administrative tasks.

The task force will need to show how the solution it proposes meets these needs, and show how it will be possible to measure progress towards achieving specified targets. The identified needs will have to be quantified. How much would be gained from quicker response to customers? How much would be gained if people spent 10 percent less time looking for information? What is the cost of not having an effective release procedure? How much would be saved if engineering changes were under control? How much would be saved if the number of administrative engineering staff could be reduced? What would be the savings from improved configuration control and traceability? What would be the benefit of synchronized engineering processes? How much would be saved by eliminating redundant data entry? How much could be saved by avoiding unnecessary transfer and conversion of data between systems? How much can be gained by improving security? What is the cost of unnecessary paperwork? What is the cost of bringing a product to market one month late? What is the cost of selling a low-quality product? What is the value of a longer sales window? What can be saved by reducing scrap and warranty costs? What is the cost of having to sell a product for more than the specified price? How much can be saved in engineering, and how much in other functions?

The other side of the coin is the investment required to implement the solution. The costs of the systems and their operation and maintenance, as well as other technical and organizational costs, should be identified. Usually, the organizational costs, are the most expensive. They are often of the same magnitude as the combined cost of hardware, software, and telecommunications. This is purely a rule-of-thumb measure, and will be very wrong in some cases. However, it serves to underline the need to consider organizational costs when calculating the total investment.

All cost-justification calculations take account of expected benefits. The problem is to understand what type of benefits should be considered, and how these should be measured. The approach taken for cost-justifying local, short-term, operational improvements is inappropriate for EDM/EWM. EDM/EWM is cross-functional and long term. Many of the benefits that are easiest to measure in a short-term project are the least important in a long-term project. A balance must be found between short-term costs, short-term benefits and long-term benefits. A short-term reduction in personnel in the "drawing store" is easy to measure, but is not the major benefit. The ability to reduce lead times by 50 percent is a major benefit, but it is difficult to measure exactly how much of this might be due to EDM/EWM, as there are so many other variables in the long term. One of the most important aspects of cost-justification is recognizing the expected benefits and appropriate measurement techniques. As the implementation of EDM/EWM is a long-term activity, the costs and benefits will need to be worked out over a five-year term. A discounted cash flow return on investment calculation will be appropriate.

It may be that the task force only proposes an initial step towards a solution, in which case a five-year approach may be inappropriate. If the task force proposes, for example, the development and use of a major prototype that will be reviewed after one year, a shorter-term approach may be suitable. However, it should be understood that the justification of EDM/EWM is a particularly complex issue. Many of the benefits are long term, and the short-term benefits are not always highly visible.

Prototyping is a short-term activity that produces useful results. However, it may appear expensive if it involves significant up-front expenditure in training and other organizational activities, and longer-term benefits are excluded from the calculation. Another factor that can distort the calculations is the discovery of existing inefficiencies in the engineering process. Most processes that are not under constant review will suffer from increasing inefficiency and waste. Improvement of process efficiency may not be considered by some to be a benefit of EDM/EWM, yet if it occurs as a result of EDM/EWM activity and produces measurable benefits, it can be regarded as a business benefit to be balanced against the cost of EDM/EWM.

It is important that the right measures of EDM/EWM performance are defined. The measures should be related to business characteristics and products, such as lead times, and not to system characteristics, such as the number of users of the system, or the volume of data managed by the system. It is not easy to identify the most suitable measures. It is just as difficult to get agreement on them from top management, the task force, and the managers who will be judged against them.

9.16 ORGANIZATIONAL ISSUES

EDM/EWM stretches across several functions and affects many people. If it is to succeed, the organizational impact will be widely felt. To minimize the negative effects and to maximize the potential benefits, a clear understanding of the organizational issues is needed. These range from basic issues, such as training, through the development of system and working procedures, to standards and policies defining the use of EDM/EWM, and major functional reorganization. The former issues, which are closer to the technology, are in general easier to address. The less technological and more purely organizational and "political" the issues appear to be, the more difficult they are to resolve. However, unless they are resolved, the benefits promised by the task force will not be achieved. Many departments will feel that the information they use is their property and their responsibility, and they will not want to see it as part of an integrated information environment in which they no longer have complete control. Similarly, they will probably overvalue the relative importance of their activities, and not want to consider themselves as part of an overall workflow.

Top management, which may have been looking for many years for ways to increase engineering productivity, may use the introduction of EDM/EWM as a lever to introduce major organizational changes. This can have a very positive effect on the introduction of EDM/EWM if it is well-managed. In other cases, it can make successful introduction very hard to achieve.

9.17 ARCHITECTURES

In the same way that an architect links the concept of a building, through plans, to the installation and use of a building, EDM/EWM architects have to be able to design and express usable EDM/EWM architectures. The cabling and piping of a building and the position of the elevators, doors, and windows must all be clearly and uniquely defined. For EDM/EWM, the key components that have to be defined include the data architecture (flow and structure), the use of data (by function and by system), the structure of the organization, the structure of control procedures (e.g. release procedures), the information flow network (e.g., sneakernet, Ethernet), the life cycle of projects and products, the computer systems involved (e.g., EDM/EWM and CAD/CAM, but also those in marketing or on the shop floor), and the structure of the EDM/EWM support organization. This information needs to be expressed in a one- or two-page, high-level summary that can be expanded to show lower levels of detail.

9.18 THE IMPLEMENTATION PLAN

The EDM/EWM strategy document defines the overall objectives, vision, scope, function, and policies of EDM/EWM, and the relationship between EDM/EWM and other activities. The accompanying overall implementation plan produced at this stage by the task force should address both the long term and the short term. For the long term, it provides management with the information necessary to understand the resources that the project will require. It shows the activities that will be required in related areas. It shows how the initial installation fits into a long-term development plan. The more specific the plan is, the better. It should define an overall implementation timetable, showing how the implementation of EDM/EWM will be split into manageable projects. For example, overall workflow control might not be implemented until the overall workflow has been optimized, yet local workflow control could be introduced earlier for some specific activities. For each project in the plan, the timing, objectives, resources, investments, relationships, and priorities should be detailed. The plan should show which people will be required to work on the projects, for how long, and when. It should show the project organizational structure. As the project will be interdepartmental, responsibilities and authority need to be clearly defined and confirmed by top management.

The short-term plan should show management which actions need to be taken in the short term. Many of these will inevitably be linked to investment in technology and organization. The plan is more likely to be accepted if it also includes some actions that will lead to short-term savings and other short-term benefits.

10

Implementation and Use of EDM/EWM

10.1 IMPLEMENTING EDM/EWM

Within the context of the overall EDM/EWM implementation activity (Exhibit 10.1), this chapter describes the second and third phases (Exhibit 10.2). These phases may still be run under the overall authority of the task force, but an increasing proportion of work and responsibility should be given to the managers and users who will actually have to use the solution that is implemented. The task force must retain responsibility for key issues, such as the implementation plan and budget, interdepartmental reorganization, training, and IT developments.

10.2 THE DETAILED IMPLEMENTATION PLAN

Chronologically, the activities should occur roughly in the order in which they are listed in Exhibit 10.2. However, many of the activities overlap, and there will

```
PHASE 1
        Preparing for Implementation
PHASE 2
        Detailed Design and Implementation
PHASE 3
        Use of EDM/EWM
```

EXHIBIT 10.1   The three-phase process

PHASE 2

> The detailed implementation plan
> Detailed architectures
> Logical data base design
> Reorganization of engineering workflow
> Technology issues
> Bringing EDM/EWM online

PHASE 3

> Detailed installation plans
> Development and purchase of hardware and software
> Integration of the solution
> Training of managers, users, and support staff
> Development of procedures and standards
> Test and acceptance of the solution
> Initial use of system
> User and manager feedback
> System use monitored, compared to expectations
> Necessary modifications made
> Maintenance and update
> Communication of success
> Progress review
> Report to management

EXHIBIT 10.2   From detailed design to use of EDM/EWM

probably be occasions where the results of a given activity or the inability to agree on a particular point will lead to previous activities being revisited.

The overall implementation plan previously produced by the task force addressed both the long term and the short term. A detailed implementation plan should now be produced addressing the short term. The more specific it is, the better. This plan should identify all necessary short-term activities. It should show which people will be required to work on the projects and for how long and when. It should show the organizational structure of projects. It should show the structure of the EDM/EWM support organization. As the projects will be interdepartmental, responsibilities and authority will have to be clearly defined and confirmed by top management.

Before EDM/EWM can be used, the EDM/EWM solution has to be built, integrated and installed. EDM/EWM architectures will have to be detailed. One important project will address data base design. Others will address the engineering workflow and engineering activities. Project life cycles and associated technical activities, systems, users, information, events, and management activities will need to be detailed. Change management procedures will have to be expressed in a logical and clear form. Users will have to be trained. Test procedures will have to be developed and put in place. Standards will have to be chosen, defined, and communicated. Detailed questions will have to be answered. How will release management really work? How will engineering and manufacturing BOMs be handled? Will it be possible to handle parametric BOMs? How will users access the system? Which users and managers should be trained first? What will the user interface look like? Where will master copies be kept? Which library symbols will be allowed? How will users be given different access rights on different projects? How will security be maintained? Will it be possible to maintain several levels of confidentiality? How will long drawings be handled by a scanner? Should the maintenance department have access to the system? Should all existing engineering drawings be scanned? How will the whole solution run on an everyday basis? What interfaces to other systems will be required?

10.3 DETAILED ARCHITECTURES

The architectures outlined in the first phase of the implementation activity now have to be expressed in more detail. They must directly reflect the chosen EDM/EWM strategy.

10.4 LOGICAL DATA BASE DESIGN

The detailed logical data base design will not be carried out by the task force itself. However, the task force will probably be responsible for this task. The task force will probably be involved in providing information for and perhaps also developing the highest levels of the design. It is an activity that will start early on, as the task force tries to understand the creation, use, and flow of engineering information, and will only be completed once the strategy has been accepted by top management, and considerable progress has already been made towards implementation. The final design will depend on the EDM/EWM strategy chosen, so some of the initial design results may eventually not be used. For example, design work on distributed data bases may prove unnecessary if the final decision is to move all data to a central data base.

As the information relationships are complex, computer-based methods must be used to document the detailing process. This does not imply that all engineers and task force members will have to use such CASE (computer aided software engineering) tools. This activity should be left to analysts and a few IT-oriented users. However, analysts cannot be expected to understand the details of engineering activities alone. Analysts, data base designers, engineers, and the task force must work together. For each major function (of both the business and the

EDM/EWM solution), the major data sets need to be identified. Data can be grouped in different ways (e.g. parts master vs. product structure, new products vs. existing products, purchasing data vs. manufacturing data). Within each data set, the individual entities need to be identified. For each file, the most important identifiers have to be identified. Exhibit 10.3 shows part of a top-level breakdown of the engineering environment. Exhibit 10.4 shows some typical data identifiers.

PROJECTS

    All Project Types
    All Project Characteristics

PROCESSES

    All Processes Used in Projects
    All Process Characteristics

PRODUCTS

    All Products
    All Deliverables for Each Product

PART IDENTIFIERS

    All Part Identifiers
    All Part Hierarchies

PART DEFINITIONS

    All Parts
    All Definitions

PART ATTRIBUTES

    All Parts
    All Attributes

TOOLS

    All Tools Used In Projects
    All Tool Characteristics

ORGANIZATION

    All Organizational Structures
    All Organizational Rules

EXHIBIT 10.3   Top-level elements of the engineering environment

Data name	Data creator
Data type	Data reviewer
Data format	Data date(s)
Data family	Data access rights
Data descriptor	Data source
Data location	Data structure
Data status	Data dependencies
Data release level	Data classification
Data owner	Data usage
Data version	Data options

EXHIBIT 10.4   Some data identifiers

Exhibit 10.5 shows specific drawing identifiers. Exhibit 10.6 shows typical part identifiers that may be found, for example, in item master/part master records. The definition data elements that can be supported need to be defined. As there is currently no real standard in this area, the solution will probably lie between the elements that IGES recognizes and those proposed with PDES/STEP. The potential states of data need to be defined (e.g., initial user development, in-process, in-review, released, under revision, withdrawn) for each data element. Data ownership issues must be resolved, and the rights and responsibilities defined for both the owners of private data and the administrators of shared data. Access, security, collection, quality, maintenance, and documentation issues must be resolved for both data and metadata. Only when the logical data base design has been completed can its physical implementation be addressed.

Part number	Date created
Drawing number	Date modified
Drawing type	Date reviewed
Drawing title	Date released
Page/sheet number	Release procedure
Revision level	Reviewer
Drawing scale	Releaser
Drawing format	Contact name
Drawing size	Standards
Drawing owner	Procedures

EXHIBIT 10.5   Some drawing identifiers

Part number	Part superseding
Part description	Part superseded
Drawing number	Make/buy source
Unit(s) of measure	Cost
Revision level/status	Lead time
Dependencies	Certifier
Creating system	Controller

EXHIBIT 10.6   Some part identifiers

10.5   REORGANIZATION OF ENGINEERING WORKFLOW

One major objective of EDM/EWM is lead time reduction. Until the detailed workflow is clear, this objective cannot be achieved. Examination of the workflow will have been one of the first activities of the task force. Its findings will serve as a starting point for a more detailed appraisal. CASE tools can be useful in this activity. When the workflow has been fully understood, it can be improved by eliminating, simplifying, or reorganizing activities. The task force will not make these detailed changes, but will be responsible for their identification and implementation. Once this activity is complete, the individual tasks, events, and control levels in the workflow will have to be defined for the EDM/EWM system.

10.6   ORGANIZATIONAL ISSUES

Successful introduction of EDM/EWM will lead to major organizational changes. Apart from modifications to individual activities and to the workflow, modifications to the organizational structure and to other control structures may also be necessary. EDM/EWM policies and working procedures will need to be defined. Users and managers from many parts of the company will have to be trained to use EDM/EWM. An EDM/EWM support team will have to be formed and trained. Its members will have to develop procedures for system management and control.

10.7   TECHNOLOGY ISSUES

Apart from the purchase and installation of new systems, it may also be necessary to carry out some in-house tailoring and development. New systems may have to be tailored so that they meet the specific requirements of the company. Interfaces may have to be developed between existing and new systems. Associated system administration and system management procedures will have to be defined. Unless the company has been very successful in the past at developing new applications, the task force should try to keep system development work to a minimum, and ensure that whatever development work does take place is kept under strict management control.

10.8 BRINGING EDM/EWM ONLINE

One of the major responsibilities of the task force at this stage is to make sure that the implementation takes place within the agreed budget and time limits. The task force is also responsible for ensuring close integration between organizational and technological activities.

10.9 USING EDM/EWM

The third phase in the process of implementing EDM/EWM is the actual use of EDM/EWM. However, in a way, this is only the beginning of the process of using EDM/EWM to produce business benefits. Before EDM/EWM can be used, the solution needs to be up and running, with new procedures in place and users trained to use them. Soon after the system is in place, users will find that it could be improved. Adaptation will take place. System use will be monitored to see that it meets expectations. Occasionally, formal reviews should take place to see what progress is being made towards meeting the targets agreed with management. Reports will be issued to management, and appropriate actions will take place.

10.10 HINTS FOR SUCCESSFUL EDM/EWM IMPLEMENTATION

From experience of various successful and unsuccessful attempts to implement EDM/EWM solutions, the following hints can be given (Exhibit 10.7).

Top management support is crucial for cross-functional, long-term projects like the implementation of EDM/EWM. Top management support for EDM/EWM must be gained and maintained, and it must be visible to all. In particular, it must be visible to middle managers who may feel threatened by EDM/EWM and unwilling to support it. It is also important that all the people who will be affected by EDM/EWM are involved with it from the beginning. This will increase their commitment to its successful use. EDM/EWM does not belong to the engineering department, neither is it a plaything of the IT department.

In any large organization, there is high resistance to change. Most people are more than willing to do tomorrow what they have been doing today. The successful introduction of EDM/EWM will require major changes in organizational behavior, and these will not be achieved through half-hearted efforts by uncommitted

Gain and maintain top management commitment.
Involve and support all functions that use engineering data.
Find and encourage champions and change-makers.
Identify and disarm opponents.
Understand business goals. Plan for business benefits.
Keep the business benefits of EDM/EWM in mind.
Plan top-down, implement bottom-up.
Keep the project plan and timetable visible.
Understand and do not ignore the organizational issues.
Develop and communicate an understandable EDM/EWM vision.
Develop and communicate an understandable EDM/EWM architecture.
Implement incrementally. Initiate pilot studies where possible.
Make sure there will be some early success.
Aim for evolutionary change in the working environment.
Aim for evolutionary product data definitions.
Remove or reorganize wasteful parts of the process.
Automate activities along the product life cycle.
Limit the number of data bases. Link activites to data bases.
Use standards where possible.
Minimize internal IT development effort.

EXHIBIT 10.7   Hints for successful EDM/EWM implementation

individuals. EDM/EWM champions, catalysts, and change-makers, in many parts of the organization, must be identified, supported, and encouraged.

The many opponents of change, whether they just prefer a quiet life or they specifically fear that EDM/EWM will reduce their power and privileges, must be identified and understood. Only then can influence be brought to modify their behavior.

The overall strategy of the company needs to be clearly understood. The only justification for the introduction of EDM/EWM is that it will improve business results. The business objectives have to be understood and quantified before the implementation of EDM/EWM can be planned. The criteria, by which the success of EDM/EWM implementation will be judged, should be defined before implementation is started. Progress towards these targets has to be monitored regularly. EDM/EWM planning should be top-down, starting from the business benefits, and working down through the workflow, towards identification of an entity's attributes. Its implementation should be bottom-up, starting from the logical data base design of an entity, and working up towards the business benefits. During planning, it is useful to identify several possible solutions, including those chosen by other companies, and understand their strengths and weaknesses. During both planning and implementation, the organizational issues should not be ignored. Often, they are as or more important than the technological issues of EDM/EWM.

Initially, people will find it hard to understand the aims of EDM/EWM. It is important to develop a clearly understandable and easy-to-communicate vision of EDM/EWM. Similarly, when the EDM/EWM architecture is defined, this should also be clear, easy to understand, and easy to communicate. If it is not, it will be difficult to enlist support from either users or managers. The task force leader must clearly understand what is being done, and be able to communicate both concepts and progress to the rest of the task force, to top management, and to users. There should be a clear project plan that is kept up-to-date and is visible to both users and top management.

Wherever possible, implementation should be incremental and based on successful pilot studies. During pilot activities, full support should be given to users. Users and their managers should be involved in the pilots as much as possible. It is much better to be able to show small but early success from real implementation than to show theoretical results and three-inch thick reports. A three-year EDM/EWM project-definition project will probably be out of date by the time its conclusions are distributed. The EDM/EWM architectures need to take account of the evolutionary nature of the engineering environment, in particular the progress that can be expected in product data definitions. Similarly, working procedures can be expected to change.

When implementing EDM/EWM, it is always useful to bear in mind that the engineering process in most companies has changed little in previous years and probably contains many wasteful activities. The implementation of EDM/EWM provides a good opportunity to identify and improve or eliminate inefficient activities that use or appear to use engineering information.

Among the major brakes on efficient use of engineering information are manual processing and manual communications. EDM/EWM implementations should aim to automate the entire process, including the communications activities. Each time a user inputs data output from one program to another, there is room for error and corresponding loss of time and money. The need to link activities to the data bases at the heart of EDM/EWM should be a key criteria for the automation of activities. The number of data bases should be kept as low as is reasonably possible. This does not imply that the aim is to have one physically centralized data base. Unconnected data base islands should not be allowed to develop around "islands of automation."

Wherever possible, standards should be used. This will increase the efficiency of engineering activity. When developing the EDM/EWM solution, international standards should be used wherever possible. This will reduce the workload both internally and when communicating with business partners. Unless it can be clearly proved that the company cannot achieve the business benefits it is looking for through the purchase of systems available on the market, in-house IT development work for EDM/EWM should be kept to the bare minimum. If it can be proved that available systems are not suitable, great caution should be exercised before proceeding with in-house developments. It may be that in-house developments are also unlikely to produce the hoped-for benefits, in which case the targets may need to be addressed through other technological or organizational approaches.

10.11 GRAND FINALE

Manufacturing companies are looking to:
- Reduce costs,
- Reduce cycle time, and
- Increase quality.

In the engineering function, manufacturing companies are looking to:

- Reduce product development costs,
- Reduce product costs,
- Reduce product development time, and
- Improve product quality.

EDM/EWM systems offer functions that can play an important role in meeting these objectives. There are also organizational ways, such as concurrent engineering, of achieving these benefits. Companies that successfully blend the technological and organizational approaches will gain the most benefits from EDM/EWM.

Appendix 1

Types of EDM/EWM Systems

A.1.1 INTRODUCTION

Currently, there are several types of systems that address the management of engineering data and workflow (Exhibit A.1.1). These range from relational data base management systems to project management systems. This appendix looks at four types of systems that are particularly relevant. They are primarily focussed on engineering drawings, engineering data in computer files, alphanumeric engineering documentation, and CAD/CAM data.

A.1.2 ENGINEERING DRAWING MANAGEMENT SYSTEMS

Engineering drawing management systems are characterized by their use of optical scanners to convert existing engineering drawings to electronic form. Exhibit A.1.2 shows some of the functions associated with these systems.

The existence of these systems is in some ways surprising, since in the 1980s many companies looked forward to a "Paperless Office," in which all engineering data would be on the CAD/CAM system. However, this dream has not yet come true. Very few companies have even 50 percent of their engineering data on CAD/CAM, and very many companies have less than 5 percent of their engineering data on CAD/CAM. As a result, they still have to deal with a vast amount of paper-based engineering information. Pushed by initiatives such as CALS or by competitive forces, they recognize the needs and benefits of digital storage and transmission of data. Since it would be both unnecessary and too expensive to enter all their engineering information in the CAD/CAM system, they take the route of document scanning.

Scanners are available for A(A4) size documents (Figure A.1.1), larger documents and for aperture cards. Paper, Mylar, vellum, and linen documents can be

Product Management System
Product Information System
Engineering Data Management System
Technical Data Management System
Technical Information Management System
Product Definition Management System
Design Management System
Engineering Drawing Management System
Engineering Document Management System
Technical Document Management System
Product Information Management System
Engineering Workflow Management System
Project Management System
Configuration Management System
Enhanced CAD/CAM Data Management System
Data Base Management System

EXHIBIT A.1.1   Some systems that address parts of EDM/EWM

scanned. Typical scanning resolutions are from 200 to 400 dots per inch (dpi), although for really high-precision work, resolutions up to 1600 dpi may be needed. Scanners work in a similar way to photocopiers and telefax machines. The raster output from a scanned drawing is in the form of a matrix of dots called a "bit-map."

A one-inch square scanned at 200 dpi will lead to a bit-map of 40,000 (200 x 200) pixels. At 400 dpi, it will lead to a bit-map of 160,000 pixels. An E(A0) size drawing (34 inch by 44 inch) scanned at 200 dpi will have a bit-map of 59,840,000 pixels. This can be stored in 7,480,000 bytes (i.e., about 7.5 MB). At 400 dpi, it

	inches	mm
A0	33 x 46.8	841 x 1189
E	34 x 44	864 x 1118
A1	23.4 x 33	594 x 841
D	22 x 34	559 x 864
A2	16.5 x 23.4	420 x 594
C	17 x 22	432 x 559
A3	11.8 x 16.5	297 x 420
B	11 x 17	279 x 432
A4	8.3 x 11.8	210 x 297
A	8.5 x 11	216 x 279

FIGURE A.1.1   Document formats

Scan and Capture Existing Drawings

Store and Retrieve Drawings

Set Up Index/Catalogue Facility

Define Attributes

Provide Secure Environment for Drawings

Manage Drawings

Archive Drawings

Enhance Drawings

Display Drawings

Edit, Mark Up, and Modify Drawings

Distribute Drawings

Print and Plot Drawings

Convert Raster Data to CAD Format

Draft

Manage the Workflow

Generate Reports

Provide System Management

Provide External Interfaces

EXHIBT A.1.2   Typical functions of an engineering drawing management system

would require about 30 MB. An A(A4) size drawing (8.5 inch by 11 inch) scanned at 200 dpi will have a bit-map of 467,500 bytes. At 400 dpi, it will need about 1.9 MB.

CCITT Group 3 and CCITT Group 4 compression techniques can reduce the storage requirements down to about 50 KB and 20 KB, respectively, for an A(A4) drawing at 200 dpi, and to about 1 MB and 200 KB, respectively, for an E(A0) drawing at 400 dpi. Equivalent A(A4) and E(A0) drawings would require about 5 KB and 35 KB in CGM vector format.

The rasterized data can be stored on magnetic or optical disks. Optical disks are WORM (Write Once Read Many times) devices. During the Write process, a high-intensity laser beam burns a hole in the disk surface, or deforms it enough to change its reflectivity. During the Read process, light from a low-intensity laser beam is reflected off the disk surface.

12-inch optical disks can hold up to 2 to 4 gigabytes. They can be used individually on a stand-alone optical disk drive. Alternatively, a multi-drive device, a "jukebox," can be used. Typically, this will hold 64 optical disk cartridges. Using Group 3 techniques, A(A4) size documents can be compressed to about 50 KB. A single-drive, 20-platter, 32 GB jukebox can hold about 100,000 E(A0) size engineering drawings.

The scanned engineering document image may need to be plotted, so most engineering document management systems include one or more plotters or printers. Typical of these are high-quality A(A4)- and B(A3)-size laser printers, A(A4)- to E(A0)-size aperture card plotters, and A(A4)- to E(A0)-size electrostatic plotters.

A local area network links the components of the engineering drawing management system together and to other local systems. Wide area networks link the system to remote locations. The "drawing manager" part of the system manages user activities, such as storing drawings, setting up a drawing index, processing queries, retrieving drawings, and communicating drawings to user workstations. The drawing manager may be based on a data base management system. Attributes such as drawing number, drawing title, revision level, and release date may be added to the basic raster information. They can then be used for queries and retrieval. The drawing manager will also have access and security features. In some systems, the scope of the metadata may be enlarged, to allow management of files, such as CAD files, on other systems.

The "process manager" allows the flow of work to be defined. It can then ensure that information is routed to the right people at the right time. View stations allow scanned images to be displayed and used. Zoom, pan, and rotate are typical features of these workstations.

A compressed raster image can be decompressed and displayed at an edit workstation. The edit function offers features such as drawing enhancement, drawing revision, and drawing redlining. Enhancement includes realigning the drawings and removing specks. Faded lines and borders can be redrawn, and text can be identified and retyped. Unwanted areas of the drawing can be removed.

Initially, these systems were primarily oriented to the input and management of raster information. Increasingly though, they offer functions to allow raster and vector information to be handled together, and to convert raster drawing entities into their vector counterparts. Raster images can be converted to vector data in a neutral format or to that of a particular CAD system. Many systems try to recognize and convert at least some entities automatically, but conversion of a typical engineering drawing usually requires quite a lot of human intervention. Most systems allow users to "draw" superimposed vector entities on raster screen entities. Some engineering drawing management systems include an integrated CAD drafting module.

A.1.3 ENGINEERING DATA MANAGEMENT SYSTEMS

Engineering data management systems primarily address digital engineering data, in particular data in files produced by CAD/CAM systems. Typically, they do not currently include an optical scanner, although this may not be so for long. Exhibit A.1.3 shows some of the functions associated with this type of system.

At the heart of the system is an "electronic vault," in which data is stored. Various types of data can be stored in the vault. The "data manager" controls access to the electronic vault, manages attribute data concerning the data, and handles queries concerning the data. The data manager is usually based on a relational data base management system.

Data can be distributed over a local area network or through a wide area network to workstations, PCs, and other systems. Large-volume data storage is available on the network. Backup and recovery facilities are provided. A report generator provides information such as project status, a history of changes, and an audit trail of attempts at unauthorized access.

The "configuration manager" maintains information about product structures and versions. It controls changes and maintains an audit trail. Different configurations, such as "as-planned" and "as-built," can be maintained for the same product. Bills of materials are maintained. "Where-used" lists can be produced.

The "process manager" allows the engineering process to be defined. As a function of events that take place, it updates the status of activities and initiates new activities. Engineering data management systems often provide interfaces to other systems, such as electronic publishing systems.

A.1.4 ENGINEERING DOCUMENT MANAGEMENT SYSTEMS

These systems primarily address alphanumeric "forms" data, although they are able to incorporate drawings as well. Exhibit A.1.4 shows some of the functions associated with these systems.

The main objective of engineering document management systems is often to communicate engineering information, such as assembly instructions, welding instructions, tooling lists and instructions, machining instructions, drawings, material lists, process plans, recipe documents, and quality control documents, to the shop floor. They can also be of use upstream, in creating specification documents, and downstream for maintenance activities. They can be linked to reference information, such as good manufacturing practices and quality system procedures and instructions.

Engineering document management systems are particularly useful for relatively

Store and Retrieve Data

Archive Data

Provide Index/Catalogue Facility

Control Data Access

Define Attributes

Define Relationships

Distribute Data

Manage Data

Maintain Audit Trail

Manage Configurations

Manage the Workflow

Manage Engineering Changes

Provide System Management

Generate Reports

Provide External Interfaces

EXHIBIT A.1.3   Typical functions of an engineering data management system

simple documents with a standard format. A template for each form is developed. Planners and other technical administration staff can then fill in the fields for a specific activity. The system can provide assistance to the planner by prompting as to likely sources of data (such as technical manuals, drawings, and other systems) and fetching data electronically when possible. Data may be input from a CAD system, a word processing system, or at the keyboard. Existing documents can be scanned into the system and stored in the data base. Indexes and cross-references can be included. Documents can be distributed around the network and accessed from workstations and shop-floor PCs.

The "document manager" controls access to documents, manages attribute information, and maintains document revision level control. Users can switch between different pages of a document, refer to cross-referenced information, and look up additional information in indexes. Some systems support hypertext features. Workflow steps can be related to documentation, thus synchronizing information flow with control of the process.

Vendors of such systems claim that they can reduce documentation generation time by 50 percent, and modification time by 90 percent. It is claimed that

Scan and Capture Existing Documents

Control Document Access

Create New Documents

Store and Retrieve Documents

Archive Documents

Display Documents

Display Graphics

Manage Documents

Modify Documents

Manage Documents Pages

Distribute Documents

Print and Plot Documents

Manage the Workflow

Provide System Management

Generate Reports

Provide External Interfaces

EXHIBIT A.1.4   Typical functions of an engineering document management system

significant gains also occur from reduction of scrap and rework, reduced storage costs, reduced operator waiting time, and improved operator understanding.

A.1.5 CAD/CAM SYSTEMS

CAD/CAM systems have always provided basic data management functions for CAD/CAM files. In some cases, such basic functions are sufficient, in others they are not. CAD/CAM vendors have addressed the problem in four different ways. Some have offered no additional data management functions. Some have offered additional functionality specific to their own systems. Some have developed a specific interface between their own system and a commercially available engineering data management system. Others have developed their own engineering data management system that can be used for data created both within their CAD/CAM system and by other systems.

MCAE (mechanical computer-aided engineering) vendors are focussing on providing tools that allow design modelling and analysis prior to CAD and hardware prototyping. These tools will help reduce development times, and can fit with DFA/DFM and simultaneous engineering approaches.

A.1.6 OTHER SYSTEMS

Developers of systems for the engineering market, other than the four types outlined above (e.g., project management systems and configuration management systems) are extending the data management functionality of their systems. They are also improving the interfaces between their own systems and EDM/EWM systems, to better address the management of engineering data and the engineering process.

Appendix 2

EDM/EWM System Overview

This appendix is based on information provided by vendors of the products and services mentioned. The products and services have not been analyzed, reviewed, rated, or checked by the author. No representation is made by the author concerning the quality, capabilities, performance, or any other aspects of the products or services described in this appendix.

It is suggested that references, documentation, and/or demonstrations be obtained from the vendor before a decision is made to purchase any of the products or services mentioned. In rapidly developing fields such as EDM/EWM, it should be remembered that market, product, and vendor characteristics can change frequently.

The intention has not been to judge or to compare systems, but to provide information on systems. The information is provided in alphabetical order by vendor. As well as engineering data, drawing and document management systems, and extensions to CAD/CAM systems, some data base management systems are also mentioned. Technical publishing systems are included, as they are major users of engineering data. Other systems, including those that address activities in a mixed raster and vector environment, are also described.

Access Corp.
1011 Glendale-Milford Road
Cincinatti, OH 45125

Product name(s):
Technical Document Management System (TDMS)
Document Filing System (DFS)
Engineering Document Image Control System (EDICS)
Image File Server (IFS)
Electronic Change Control (ECC)

TDMS is ACCESS's integrated hardware/software solution for document

157

management. TDMS is designed as a workflow engine, providing online access to document and data files as information relates to these documents and files at any time during their life cycle. A full TDMS includes at least one IFS, EDICS (running on a host computer), input and output devices (scanners, printers, etc.) and network communications capabilities.

EDICS is one of the subsystem options of a TDMS. EDICS provides engineering and manufacturing document management capabilities. EDICS can manage documents ranging in size from 8.5 inches by 11 inches, or smaller, to documents as large as roll drawings 44 inches wide. EDICS uses the ORACLE relational data base to manage information about documents. When integrated with an IFS, EDICS offers document-object (image and data files) management capabilities. EDICS runs on a DEC VAX, using the VMS operating system.

ECC is an EDICS option that offers Approval Processing, Electronic Sign-off/Rejection, Approval Status Reporting, Approval History Reporting, Alert Processing, and Automatic Distribution. ECC automates the engineering and technical document approval process consistent with each organization's unique document approval process. During this process, each document undergoing approval processing is assigned an approval state or condition. These states include approval in process, rejected, approved, and incorporated. ECC approval processing can be performed on any document type.

The IFS is a 80386-based computer that runs the Document Filing System under the Unix operating system. The IFS provides shared access to document-objects (image and data files) stored in optical and magnetic disks.

TDMS has five subsystems. The Control Subsystem includes the hardware and software components related to EDICS, or a customer host computer that provides functionality similar to EDICS. The Storage Subsystem includes the hardware and software components related to the IFS and its associated optical and magnetic storage. In a stand-alone IFS configuration, the Storage Subsystem also provides limited Control Subsystem functionality. The Input Subsystem includes the hardware and software used for the input of document-objects (images and data files) to the TDMS, as well as the information related to these files (document descriptors). Scanners, QC workstations, and the CAD File Manager (when used for TDMS input) are all considered components of the Input Subsystem. The Output Subsystem includes the hardware and software used for the output of document-object (images and data files), as well as the information related to these files (document descriptors) from the TDMS. VIEW and REDLINE workstations, printer controllers, printers, and plotters are all considered components of an Output Subsystem. The Communication Subsystem includes the hardware and software components related to intersystem and TDMS network communications. Such hardware and software would include the networking cabling, terminal server(s), IFS Network Processor(s), the Network Management Console, and so on. ACCESS manufactures a multiple platter jukebox capable of storing up to 40 GB of data.

ACCESS supports the storage and management of files in all CALS formats, including MIL-D-28000, MIL-D-28001, MIL-D-28002, MIL-D-28003.

Users of ACCESS's integrated technical document management systems include McDonnell Douglas Space Systems Company, Douglas Aircraft Company, Hughes Aircraft Electro-Optical and Data Systems Division, Burndy Corporation, Compaq Corporation, Martin Marietta Energy Systems, TRW, and Ford Motor Company.

ACS Telecom
25825 Eshelman Avenue
Lomita, CA 90717
Product names(s):
Personal EDMS
AutoEDMS
Personal EDMS is drawing management software for individual AutoCAD users. AutoEDMS is a data management system for networked AutoCAD users in Novell Advanced NetWare 286 and 386, and SUN PC/NFS environments. AutoEDMS links AutoCAD drawings, related data, and raster images, and provides group, sort, search, load, view, and revision tracking functions.

ADI Software GmbH
Rheinstrasse 122
7500 Karlsruhe 21
Germany
Product name(s):
ADICAD
ADICAD is a data management system for Hewlett Packard's ME 10 and ME 30 CAD systems.

Adra Systems Inc.
59 Technology Drive
Lowell, MA 01851
Product name(s): Adra Vault
Adra Systems, Inc. manufactures and markets workstation and PC-based CAD/CAM solutions for mechanical design and drafting, numerical control, and engineering drawing management applications. Adra's CADRA-III design/drafting software, CADRA-NC multi-axis NC programming package, and Desktop Prove-Out tool-path verification package are available on performance-optimized Adra workstations and standard platforms, including those manufactured by Digital, Sun, Tektronix, and Apollo. Since 1985, Adra has delivered over 3500 systems

to customers in the aerospace, automotive, metalworking, office products, plastics, and defense-related industries.

The Adra Vault is a multi-functional drawing management system that allows Adra customers to manage drawings and related documentation. The Vault controls access to documents, maintains an orderly release process, and manages information about what is in the drawings or documents. It handles CADRA-III drawings, as well as files from other CAD systems, and any other file that needs to be controlled. The Vault manages both standard and user-defined attribute data that may be taken directly from drawings or input by the user.

Versions of the Vault run on DEC VAX computers, Sun Microsystems, other Unix-based systems. The Vault has three components:

- A relational data base management system (EMPRESS), running on the host,
- The Vault drawing data base management software, residing on the host, which maintains the Vault data base. This permits data input, ad-hoc queries, archiving, report generation, sign in, sign out, and system manager functions, and
- For users of CADRA-III CAD/CAM systems, the CADRA-III software, which creates drawings, checks them in and out, and defines attribute data to be entered into the relational data base.

For very large applications, the Vault may be distributed over NFS or DECnet networks. In these environments, the EMPRESS data base used by the Vault is fully distributed over the network.

The Vault stores and manages files that describe objects to be manufactured by the organization. The files stored in the Vault may be CAD files from either CADRA-III or other CAD systems, or they may be files of other types, such as NC programs or administrative data. The Vault manages multiple copies of any type of file.

Users sign drawings and files in and out of the Vault. When sign-outs are done, the Vault checks users privileges and whether another user may have already signed the drawing out. Each user has specific privileges, and these are checked on all operations.

The Vault tracks revisions of drawings as they proceed from engineering through release. Each revision of a drawing has ECO information, and data is kept on whether a drawing is being edited, has been signed back in, has been checked, and has been released. Promotion activities, such as checking or releasing a drawing are privileged operations and can only be done by users with those specific privileges for a specific project.

The Vault provides for complete bill of materials, expanded bill of materials, and where-used capability. Product structure (i.e., the assembly, subassembly, and component hierarchy) can be driven directly from CADRA-III drawing files, or can be manually input through the menu interface. There is no limit to the depth or complexity of the bill of materials tree. It may cover as many levels of assembly, subassembly, and components as required, and may reference standard parts or components that were defined for different projects.

Alcatel CGA-HBS
Division Informatisation Administrative
B.P. 57
Le Plessis Paté
91220 Brétigny-sur-Orge
France

Alcatel CGA-HBS' "SERVOPLAN" system manages engineering documents.

Alcatel TITN
1, rue Galvani
BP 110
91301 Massy Cedex
France

DELTOS, Alcatel TITN's technical document management system, has seven principal modules:

- Document management (raster files, CAD files, electronic documents, etc.),
- Document storage (on optical disk),
- Document search, retrieval, and view,
- Compound document publication,
- Rasterisation (CCITT Group IV compression),
- Conversion (standard formats),
- Communication (print/plot and over networks).

DELTOS uses the SYBASE relational data base, and runs on Sun workstations.

Aldus Corporation
411 First Avenue South
Seattle, WA 98104

Aldus Corp.'s PageMaker is a leading electronic publishing program.

Alpharel Inc.
3601 Calle Tecate
Camarillo, CA 93010

Product name(s): Digital Documentation Transmission and Management (DDTMS)
Since incorporating in 1981, Alpharel has developed and delivered medium- to large-scale document management systems. These modular systems include Alpharel's hardware, software, and communications components, as well as equipment and software from other vendors that incorporate scanning, storage, data base, networking, and communications technology. Alpharel's products are designed for IBM host computers of many different capacities. These products

provide organizations with the "building blocks" to construct the flexible, application-specific systems they need for efficient, computerized management of both small- and large-scale document archives.

The seven major modules are :

- Document scanning,
- Document storage,
- Document retrieval,
- View and print,
- Communications,
- Management tools, and
- System utilities (e.g., export/import images from tape).

The document management capabilities can be provided either with a central host system, where documents and the data base are co-resident, or with a distributed system that has connections to a central host for access to archived or remote documents, or to a central data base.

DDTMS maintains close control of each drawing revision, restricting changes in revisions to authorized users. All drawings and users are classified by predefined authorization tables. Alpharel uses IBM's relational data base management system DB2 for the drawing index and for storage of the image "object."

Alpharel's software is also capable of functioning with IBM's Data Communication Service (DCS) facility. DDTMS uses DCS to manage the data generated by multiple applications in multiple environments. DCS stores data in a relational data base (DB2, for example) that can be either centralized or distributed. In this manner, raster images scanned in and stored by DDTMS along with vector models from various sources can be stored in the same central repository and accessed by any authorized user.

Alpharel's image management systems provide for complete data integrity and security. These provisions range from internal and external security checks to backup and recovery procedures that prevent both data loss and degradation, as well as unauthorized document access.

Alpharel designed and installed the largest optical disk-based engineering drawing management system in operation for the U.S. Army and Air Force. Alpharel also counts among its customers organizations like Pacific Gas and Electric, Ford Motor Company, Lockheed Aeronautical Systems Company, General Electric Nuclear Energy Business, Apple Computer, Inc., General Electric Aerospace, and the U.S. Navy.

Archivage Systèmes
Zone Industriel du Vernis
29200 Brest
France

Gédéon is an optical disk-based drawing and CAD data management system for UNIX and DOS environments. Based on Ingres and Hypertext modules, it allows users to scan, view, store, organize, index, manage, and maintain both text and graphics.

Elf Aquitaine is a Gédéon user.

Aries Technology, Inc.
600 Suffolk Street
Lowell, MA 01854

Aries Technology is a leading supplier of MCAE systems.

ASL
Denbigh House, Denbigh Road
Bletchley, Milton Keynes MK1 1YP, England

Product name(s):
Drawing Office Records Information System (DORIS)
Engineering Request for Information Change (ERIC)

DORIS has been developed to control records to BS 5750/MOD 05-10 standard. Written in Dbase III/Clipper, it can be used in a single or multi-user environment. It provides Document Register, Parts and Item List Register, Build Standard Register, and CAD Control capabilities. Status, tracking, reporting, and history functions are provided. With ERIC, change authorization rights are managed, and change notes and deviation notes can be handled.

Autodesk Inc.
2320 Marinship Way
Sausalito, CA 94965

Autodesk Inc. is the developer of AutoCAD, the leading PC-based CAD system.

Auto-Scan Systems Inc.
7 Piedmont Center
Suite 408
Atlanta, GA 30305

RETREEVE is an Oracle-based drawing management system that allows users to index, store, locate, view, and print raster and AutoCAD files. Display and printing of HPGL files are also supported. RETREEVE works in stand-alone and networked environments.

Auto-trol Technology Corporation
12500 N. Washington Street
Denver, CO 80241

Product name(s):

Technical Information Management (TIM)
Engineering Information Management System (EIMS)

Auto-trol Technology Corporation was founded more than 25 years ago, as a builder of digitizers and flatbed plotters. In the 1970s and 1980s, it was one of the companies to pioneer new CAD/CAM technologies.

To address engineering data and document management needs, Auto-trol has introduced TIM, a series of products and integrated support services that expands traditional CAD/CAM capabilities. TIM helps manage the engineering process from idea through implementation, using state of the art tools for scanning, imaging, storage, distribution, and relational data base management, combined with computer-aided graphic solutions for specific engineering applications. For example, the Scanner Subsystem is used to scan and clean up hard copy drawings using a variety of scanners. The Composite Image subsystem allows users to introduce scanned images into their CAD system, combine them with vector entities, update them electronically, or use them as the basis for new drawings. The PreVIEW/PrePARE products provide electronic methods for distribution, review and markup of drawings via PCs, or engineering workstations.

EIMS is the central element in managing the information data base. It serves as the "electronic librarian" that tracks both graphic and non-graphic engineering information stored on-line or in CAD archives. It assures that the latest versions of drawings are always accessible and available immediately for use in current or future projects. It shows where drawings are stored, who uses them, how often, and in what projects. User privileges can be assigned to maintain security of confidential files without hindering system flexibility or ease of access to other material.

The data base is accessed and controlled either through a menu or command language. The menu provides prompts and a listing of options, so casual operators can use the system effectively. The EIMS command language allows more advanced users to enter instructions directly. A structured query language (SQL) also provides immediate access to the data base. EIMS includes a menu-driven capability for generating formatted reports on project status and system operation.

TIM solutions for archival, information scanning, remote drawing review and annotation, optical storage, and data base management can be integrated with Auto-trol's Tech Illustrator and Tech Image application tools, to address automation tasks for new and existing documents. Solutions are available to meet needs for engineering and manufacturing documentation, particularly in the areas of illustrated engineering sales proposals, engineering standards documents, manufacturing assembly documentation, parts and service publications, and other technical publishing requirements.

AutoSIGHT Inc.

P.O. Box 362086
Melbourne, FL 32936-2086

Product name(s):

AutoSIGHT
AutoSIGHT CONVERT
AutoSIGHT COMPARE
AutoSIGHT MINI

AutoSIGHT features include file locking, layer control, file conversion, data extract, redlining, overlay, multiple format viewing, and so forth.

AutoSIGHT CONVERT provides file conversion capability to and from file formats, such as DWG, DXF, and HPGL. AutoSIGHT COMPARE lets users view differences between drawings in DWG and DWF formats. AutoSIGHT MINI is a compact view-only file access tool.

Boeing Computer Services

P.O. Box 24346
Seattle, WA 98124-0346

Boeing Computer Services' PPDM (Product and Process Document Management) consists of core facilities, application modules, and custom system interfaces.

The core facilities are:

- System administration,
- Application interface controller,
- Document management (PPDM-controlled repository), and
- Information management (messages).

The application modules are:

- Work authorization,
- Configuration management,
- Project management and event administration,
- Release management,
- Document management (externally-controlled repository), and
- Archive management.

Typical custom system interfaces are:

- CAD/CAM,
- Scanners and optical disk storage,
- Facsimile,
- Workstations, printers, and plotters,
- Publishing systems,
- MRP 2 systems,
- Shop floor control systems, and

- Device management software.

Boeing Computer Services' PPDM was designed to work in the Tandem Computer environment. Boeing Commercial Airplane is a user of PPDM.

CADCO Ltd.
Penrose House
Penrose Quay
Cork, Eire

CADCO markets the CADMAN drawing management system developed by SMC Ltd. CADMAN provides a management shell around AutoCAD. As well as offering access and password protection, it controls and records drawing movements, and provides extensive search and retrieval capabilities. Single user and networked versions are available.

CADKEY Inc.
440 Oakland Street
Manchester, CT 06040

CADKEY is a leading PC-based CAD system.

Cadre Technologies Inc.
19545 NW Von Neumann Drive
Beaverton, OR 97075

Product name(s): Teamwork

Cadre Technologies Inc. develops, markets, and supports integrated software and hardware solutions and tools to manage and automate the design, code generation, and test of high performance computer systems and software. The company's three divisions—the Teamwork Division, the MicroCASE Division, and the Atron Division—support a fully integrated Computer-Aided Software Engineering or "CASE" environment. The three phases of software development covered by Cadre's product lines include the design definition and analysis prior to programming, code generation, and the verification of the completed code's performance with the target system.

Cadre's products provide comprehensive coverage from system design capture through integration and test. Systems analysis, design, simulation, source code generators, language tools, debuggers, performance analysis, and verification are each addressed through one or more of the company's products.

The Teamwork Division provides integrated front-end tools for systems development through its Teamwork-ADAS product set. The core of the Teamwork toolset automates structured software engineering methods using interactive computer graphics and multi-user workstation power strategically positioned for heterogeneous networks. Teamwork runs on standard workstation platforms from Apollo, DEC, HP, IBM, and Sun Microsystems.

CADworks Inc.
222 Third Street
Cambridge, MA 02142

CADworks develops the DRAWBASE family of PC-CAD software. This offers integrated two-dimensional drafting, three-dimensional design, and data base management capabilities. CADworks has incorporated Image Systems Technology's CAD Overlay in DRAWBASE, to help users with the conversion of scanned hardcopy drawings to vector format. There are more than 3500 DRAWBASE installations worldwide.

CAYLX Software Ltd.
15245 Pacific Highway South
P.O. Box 69049
Seattle, WA 98168

CAYLX was incorporated in 1985. With PC-based TIPS (The Image Processing Software), users can scan in documents, select required parts, and integrate these document images into existing computer applications on mainframes, minis, and LAN servers.

Cimage Corporation
3885 Research Park Drive
Ann Arbor, MI 48108

Product name(s): ImageMaster

Cimage Corporation is a supplier of Technical Document Image Management Systems (TDIMS) for small and large format technical documents, running on industry standard PC and workstation platforms under UNIX and DOS. The Cimage ImageMaster product provides a document management and distribution system for technical documents, including engineering drawings, process sheets, product assembly documents, and so on.

ImageMaster scans existing paper and microfilm documents, and can accept drawings directly from CAD systems. It also provides a document archive that can be accessed over local and wide area networks. Document Management features include redline management, check-in/out, revision handling, document linking, document structuring, tailorable indexing, and a tailorable user interface. Document distribution may be controlled by the ImageMaster Document Manager, or may be transferred to mainframe hosts for archiving and centrally controlled distribution.

ImageMaster provides a range of workstation functions, ranging from simple search and retrieve, to comprehensive raster, 2-D vector, and full text editing capabilities in a fully integrated graphical editor. ImageMaster workstations can be networked together with local ImageMaster Document Management Systems, to provide technical document solutions.

Cimage customers include British Aerospace, British Petroleum, British

Telecom, Fiat-Iveco, Ford Motor Company, General Motors Corporation, and Statoil.

Cimlinc
700 Nicholas Boulevard
Elk Grove Village, IL 60007

Product name(s): Intelligent Documentation

Cimlinc's Intelligent Documentation is an automated document generation, transmission, and maintenance program that can be integrated with a variety of CAD systems. It retrieves and integrates text and graphic information stored in networked homogeneous computing environments, to generate production-ready documents for the shop floor. It can be run independently of Cimlinc's Cimcad system. The system, running on EMPRESS data base management software, is in use at Boeing Aerospace, Cincinnati Milacron, General Motors, Harley Davidson, John Deere, Martin Marietta, and Westinghouse.

Compugraphic Corporation
200 Ballardvale Street
Wilmington, MA 01887

Compugraphic's Computer Automated Publishing System (CAPS) is a leading technical publishing system.

Context Corporation
8285 S.W. Nimbus Avenue
Beaverton, OR 97005

Mentor Graphics Corporation, parent company of Context Corporation, is a leader in electronic design automation (EDA).

  Context's Documentation Management Environment (DME) includes:

- Structured documents,
- Distributed document architecture,
- Customization,
- Change control,
- Concurrent authoring and review,
- Open architecture,
- Process modelling and management,
- Configuration management,
- Viewing management, and
- Navigation management.

  Context Corp. also markets technical publishing software.

Control Data Corp.
8100 34th Avenue South
Minneapolis, MI 55440

Product name(s): EDL

EDL is Control Data's Product Information System (PIM) application. Initial design efforts for EDL began in 1979, with the production release of the product occurring in 1982. EDL is an application-independent networked information system. It supports the tracking and management of electronic information in a distributed, heterogeneous environment. EDL fully integrates the Control Data ICEM suite of design, analysis, and manufacturing products into a complete product design solution. The EDL Integrator supplies utilities and tools for the incorporation of existing site-specific applications into the EDL umbrella.

EDL can conform to the existing manual processes and procedures used by a company. EDL provides the flexibility to ensure an implementation is user friendly and responsive at all levels of a physically distributed, logically integrated environment. Using the utilities and tools, users can create a tailored environment and system that reflects their unique operational and business requirements.

EDL applies computer-based technology integrated with information management technology to solve problems in using and managing distributed environments. EDL embraces a two-part philosophy that aims to eliminate the complexity of operation and computerese required of users to perform their tasks, and to create a logical computing and information resource that spans a heterogeneous, distributed environment.

The EDL strategy is to capture information about data at its time and point of creation, eliminating redundant and possibly erroneous data entry. EDL can manage and control access to any type of application and the information created by the application, and control the processes for the distribution and review of such information. EDL's information distribution and sharing capabilities support simultaneous engineering.

EDL supports a repository of information about products, part structures, and design information (geometry, analyzes, detail drawings, etc.) that can interface to other corporate systems, such as MRP 2, quality assurance, and financial planning. EDL provides the following capabilities:

- User environment creation and management,
- Application interface and management,
- Common user interface across systems and functions,
- Design/data capture and classification, tracking and retrieval,
- Full product tailorability,
- Electronic review and release processes,
- Access management and control,

- Distributed administration tools,
- Backup and archive,
- Multiple, distributed data repositories,
- Network access and transfer (batch and interactive),
- User and system administration,
- Query and reporting, and
- Installation, customization, and implementation utilities.

These capabilities are provided in a distributed environment of heterogeneous systems, including Control Data mainframes and workstations, Sun, Apollo, DEC Ultrix, IBM RS6000 and Silicon Graphics workstations, and 80386 Unix-based PCs.

EDL is built on relational data base technology and supports multiple shared data bases, hot backup of active data bases, and organizational hierarchies of data bases. EDL is delivered as a set of functional modules, such as EDL Networked Information Manager, EDL Networked Administrator, and EDL Integrator. The EDL Release Manager module provides for defining and managing review and release functions. The EDL Product Structure Controller module provides for creation, control, and maintenance of product structures.

EDL is installed at over 200 sites worldwide. These sites include companies in the automotive, aerospace, defense, heavy machinery, and other manufacturing industries.

Cyco International
1908 Cliff Valley Way
Suite 2000
Atlanta, GA 30329

Product name(s):

AutoBASE
AutoManager

AutoManager allows users to organize, locate, view, and manipulate DWG files. In addition to these functions, AutoBASE offers data base functions for drawing management, indexing, storage, and protection.

C-TAD Systems Inc.
Atrium Office Center
900 Victors Way
Ann Arbor, MI 48108

Product name(s): The Integrator

C-TAD, incorporated in 1983, develops the "Integrator" line of direct CAD data translation products.

Dassault Systèmes
24-28 ave Général de Gaulle
92156 Suresnes, France

Product name(s): Catia Data Management (CDM)

Dassault Systèmes was set up in 1981 to further develop and market the CATIA CAD/CAM system developed by Avions Marcel Dassault (AMD-BA). More than 1000 companies use CATIA. In addition to continuing the development of CATIA, Dassault Systèmes has developed complementary products, such as Catia Data Management (CDM).

CDM is designed to integrate and manage both CATIA and non-CATIA data. It runs in the IBM environment on the relational data bases SQL/DS (VM) and DB2 (MVS). CDM provides secure, shared data access, as well as revision control, data management, data extraction, data structuring, and data distribution features. The CDM-A (Catia Data Management Access) module allows CDM to be used from a CATIA graphic screen.

Database Applications
14 Admiralty Place
Redwood City, CA 94065

Product name(s): CADEX

Database Applications' VAX-based CADEX is a computer-aided card catalogue. It offers document management and information sharing for projects, departments, and companies.

Digital Equipment Corporation
4 Results Way
Marlboro, MA 01752 9103

Product name(s): Electronic Data Control System (EDCS-II)

EDCS-II is an integration tool that provides tracking, access control, change notification and archiving of data across a network regardless of the application that was used to generate the data. It can manage the data files produced by a heterogeneous group of applications, such as CAD/CAM, NC, sampling, test, and maintenance programs. It can also manage nonelectronic data and documents, such as design drawings stored on microfilm or hardcopy. EDCS-II manages important information (metadata) concerning the documents it is controlling. It enables users to efficiently share data, protects data from unauthorized access, and ensures data integrity.

EDCS-II provides revision tracking of work in process, work in review, and released documents. It provides an electronic notification capability that informs users of changes that may affect them, and informs them of current status. EDCS-II also provides configuration management and version control features, data base querying capabilities, and a configurable review manager. EDCS-II offers four types of interface, including DECwindows menus and a callable interface.

EDCS, the forerunner of EDCS-II, was launched in 1986. EDCS-II was introduced in late 1989. Compared to EDCS, its major additional features are configuration management and customizable review management.

Users include Hughes Aircraft Electro-Optical and Data Systems Group, Lockheed Sanders, National Semiconductor, Navistar, Alcatel AFTH, Robert Bosch, Selenia, and Siemens.

Docugraphix
1601 Saratoga Sunnyvale Road
Cupertino, CA 95014

Docugraphix is a leading developer of engineering document management systems.

Eigner + Partner GmbH
Ruschgraben 133
7500 Karlsruhe 1
Germany

EP-SDV is the technical data base management module at the heart of the CADIM (Computer-Aided Data Integration Manager) information system.

E. I. duPont de Nemours & Co.
Imaging Systems Department
Eagle Run
Newark, DE 19714-6099

DuPont's FastTrax System is a raster-based engineering drawing management system. It consists of a stand-alone or networked group of Macintosh IIx platforms, a direct thermal plotter, an optical storage system, and FastTrax Management software. The Scan Station can be used with a variety of scanners.

Luz Construction Inc. uses FastTrax.

Electronic Data Systems Corporation
1400 North Woodward Avenue
Bloomfield Hills, MI 48013

Electronic Data Systems Corporation (EDS) has developed the Technical Data Management Facility (TDMF). TDMF automates the management, control, and distribution of information in multi-media format.

Based on the incorporation of components, TDMF provides a set of tools to help manage, control, and distribute valuable resources and information. This "open systems" architecture gives the freedom to add modules as required. System components are selected to satisfy the many varying requirements for performance and cost, and provide a wide range of functions.

Through the TDMF system, access to the EDS private data network provides users with both local and wide area networks. The information distribution system of the TDMF Global Data Network offers worldwide sharing and a variety of applications and data information through a single point of access.

The selection of hardware components available includes the following:

Processors and Operating Systems—Mainframes: IBM, MVS, Minicomputers: DEC, VMS

Access Devices—Personal computers, intelligent workstations, graphic consoles, and ASCII terminals

Storage Devices—Magnetic tape, magnetic discs, and optical media

Input Devices—Paper scanners (page and large format), aperture card scanners, and CAD systems (vector and plot files)

Output Devices—Electrostatic printers, laser printers, dot matrix printers, aperture card reproducers, and optical media, including CD-ROM

The TDMF system can be customized to fit unique or site-specific procedures that must be incorporated into system design. This customization encompasses both the user interfaces and TDMF functions, and provides the ability to access information from within applications, while maintaining a standardized function across all platforms.

The tailored TDMF system integrates multiple levels of information and provides easily accessible data. The system provides data control, distribution, translation invocation, change notification, and system security.

Eyring Inc.
1455 W. 820 North
Provo, UT 84601

Product name(s): Impression

The Impression system provides for the creation of on-screen documentation, complete with indexing and cross-referencing. It incorporates graphic images with text into work instructions and displays them at the point of assembly. Color highlights and labels can be added to coordinate the text with the graphics. Easily updated instructions are displayed almost immediately to low-cost PC workstations.

Impression software links work scheduling and display of work instructions with the reporting and storing of production data (e.g., standard or nonstandard completion, test suites, assembly time). The software allows production and operations personnel to use essential information from and feed information to diverse plant sub-systems, without requiring the operator to leave the Impression environment. This ensures that information is routed in the most direct manner possible. In addition, the software allows execution of external programs at the workstation level with automatic reentry into the work instruction flow.

Instruction Authoring is the Impression subsystem for combining text and images and adding color highlights. Images can range from line drawings to black and white photographs. Instruction Authoring includes the ability to assign cross-references to any page and to index the data base on related topics. It makes it possible to maintain an "official copy," the copy the worker sees, and a "working copy" on which changes can be made. As soon as changes are finalized and approved, an authorization step allows the authorizer to change the status of the working copy to the official copy. The newly authorized copy then becomes available to the shop floor worker immediately.

The Instruction Display subsystem allows the shop floor operator to view pages of work instructions, call up cross-referenced standards or procedures, and look up indexed information. Detailed work instructions are displayed at a local workstation within moments of an operator's request. Only authorized work instructions are available to the shop floor operator.

Impression's data base contains a catalogue of images stored in highly compressed from. The image processing function provides the means of acquiring and storing images for use in creating work instructions. Images can be scanned or converted from a digital format. Once converted, the images are catalogued for use by authors in the assembly of work instructions. The image processing tools include scanning, image file conversion, and image cataloguing.

The Workstation Manager Module links Impression control to the entire range of factory automation operations. The Workstation Manager coordinates a user's interaction with different programs at the workstation level. This ensures that information is presented to the operator and collected from that workstation in the most direct manner possible.

Eyring Inc. was founded in 1972.

FileNet Corporation
3565 Harbor Boulevard
Costa Mesa, CA 92626

Since 1982, FileNet has developed and marketed document image processors and associated integrated document storage and retrieval systems. It is a market leader in the number of systems installed to automate paper-intensive business applications. WorkFlo is Filenet's workflow application.

FORMTEK, Inc.
661 Andersen Drive
Pittsburgh, PA 15220

FORMTEK Inc., a Lockheed company, was founded in 1982. FORMTEK supplies image-based engineering document management systems. FORMTEK's software solutions include:

- FORMTEK:TIMS—Document archival and retrieval,
- FORMTEK:SCAN—Capture of engineering documents,
- FORMTEK:CLEANUP—Document enhancement,
- FORMTEK:CONVERT—Image file conversion,
- LaserSystem/UX—Optical disk manager,
- LaserStar/UX—Optical jukebox manager,
- FORMTEK:REDLINE—Electronic mark-up on workstations,
- FORMTEK:REDLINE.PC—Electronic mark-up on PCs,
- FORMTEK:SKETCH+—Raster editing,
- FORMTEK:DRAFT+—ANSI-compliant overlay drafting,
- FORMTEK:NET—Object connectivity for network applications,
- FORMTEK:VIEW—Immediate document access for authorized users,
- FORMTEK:VIEWPC—Workstation viewing capabilities on a PC,
- FORMTEK:PLOT—Plot of paper documents/aperture cards, and
- FORMTEK:FAX—Large document faxing.

FORMTEK products run on Sun Microsystems' family of UNIX-based workstations, DEC's family of VMS workstations, and IBM's family of AIX workstations. Components of the FORMTEK system are connected over Ethernet.

FORMTEK's customers include Bell of Pennsylvania, Boeing, GE Aircraft Engines, Pratt & Whitney, Pacific Bell, U.S. Defense Logistics Agency and U.S. Navy (for EDMICS), Rolls-Royce, Short Brothers, and RTT (Belgium).

GTX Corporation
8836 North 23rd Avenue
Phoenix, AZ 85021

GTX Corporation develops document management and CAD conversion systems. It offers a line of products designed to stop drawing deterioration and improve the quality of drawings over time. Founded in 1984, by 1989 it had shipped over 125 systems. Customers include Toyota Motors and the U.S. Navy.

GTX Conversion Series products offer fast, accurate raster-to-vector conversion. The Expert Series is designed for a high-volume production environment, where greater throughput of drawings to CAD is a necessity. Input and output files use industry-standard formats. An output CAD file includes standard entity types (lines, arcs and circles, filled areas, etc.), symbols, and text. Entities are sorted onto layers. The Professional Series is designed for the smaller site, where throughput speed is not so critical, but accuracy is a must. The GTX Document Management Series products provide control over drawings in raster format. This includes complete raster editing, viewing, markup, and distribution control.

GTX modules include:

- D-RAST Mark-up,
- D-VIEW View, Mark-up, Query, Retrieve,

- D-FILE Drawing control under ORACLE,
- D-CAD Vector CAD editor,
- D-EDIT Raster CAD editor, and
- REMOD Scan, Enhance, Index, Compress.

All GTX products accept and output the raster and vector CALS formats specified for engineering drawings.

Gerber Systems Technology Inc.
425 Sullivan Avenue
South Windsor CT 06074

Gerber Systems Technology Inc., a subsidiary of Gerber Scientific Inc., develops the SABRE 5000 CAD/CAM/CIM system for mechanical design, NC, and CIM applications. Gerber's OASSIS is an electronic shop documentation system that combines form creation, text editing, and graphics capabilities. Existing forms and drawings can be scanned in or imported from a CAD system.

Hewlett Packard
Mechanical Design Division
3404 East Harmony Road
Fort Collins, CO 80525-9599

Product name: HP Data Management System (HP-DMS)

HP-DMS is designed to help the CAD designer manage, administer, and control design information. It is offered as an optional part of the three ME CAD products from HP: the ME Series 10 Design and Drafting System, the ME Series 30 3-D Modelling System, and the ME Series 10v View and Plot System. HP-DMS is based on the SQL data base query language, and supports relational data bases from HP (ALLBASE/SQL and SQL300) and from Oracle Corporation.

Features of HP-DMS include:

- Management of drawing and three-dimensinal model versions and revisions,
- Full data backup and archival capabilities,
- Access control mechanism to ensure data integrity,
- Title block management providing link between drawing and database information,
- Full multi-level BOM capabilities,
- Company-defined part classification system,
- Support for heterogeneous (CAD and non-CAD) data,
- Transparent user interaction with network file systems, and
- Fully customizable user interface.

HP-DMS is currently positioned as a CAD department level solution in the electronics, electro-mechanical, machine tool, and fabricated metal parts industries. Future product development will be aimed at broadening the range of engineering disciplines (in particular, EE CAD) and supporting site/division level requirements, such as tight integration with manufacturing.

Through HP's Value-Added Business program, HP works with Value-Added partners, supplying Product Information Systems that run on HP computer platforms. In many cases, these solutions complement HP's own product HP-DMS.

Index Technology Corporation
One Main Street
Cambridge, MA 02142

Index Technology develops the EXCELERATOR Series Computer-Aided Software Engineering (CASE) products.

Informix Software, Inc.
4100 Bohannon Drive
Menlo Park, CA 94025

Informix develops data base products. Informix-SQL is a relational data base management system.

Intergraph Corporation
One Madison Industrial Park
Huntsville, AL 35807

Product name(s):
DMANDS (Drawing Management and Distribution System)
I/PDM (Intergraph/Product Data Manager)

DMANDS is a drawing conversion, management, maintenance, and distribution system for large-volume drawing libraries. The I/SCAN module works with a variety of scanners. I/RAS 32 offers raster display, cleanup, and editing capabilities. View Only and Redline software provide viewing and markup facilities. I/VEC converts raster data to vector data. I/SCR processes symbols and characters identified by I/VEC. I/IMAGE raster and vector software can be used to view and manipulate grayscale and 24-bit color images.

DP/Publisher is Intergraph's technical publication software.

I/NFM (Intergraph/Network File Manager) manages files, both current and archived, in a heterogeneous or homogeneous network environment. Built on top of Intergraph's Relational Interface System (RIS), it is independent of the underlying relational data base management system. Ingres, Oracle, and Informix are supported.

The product set I/PDM-I/PDU organizes, manages, and provides easy access to information about parts and assemblies and their files in a networked environment. I/PDM, which runs on a central server machine, uses a relational data base to store part and assembly metadata, including information about the location of the files. The metadata can reside on the central server node or on any other node in the network. The files can be stored in a central node or distributed in many storage nodes in the network. I/PDU, which runs on the workstations, is the user interface to I/PDM. I/PDU is integrated with Intergraph's mechanical design products (I/EMS-I/DRAFT-I/MDS) and provides seamless access to the part and assembly data, and facilitates file movements between the file storage locations and users' workstations. I/PDU also supports assembly design activities. The relationships between parts and assemblies are stored in the relational data base; and reports such as single level and exploded bill of materials can be generated from them. Utilities for archiving/restoring part/assembly files are included in the product. Access control and electronic approval cycle mechanisms are built into the product. The I/MRP product enables the extraction of part/assembly information from the PDM database, and formats them for input to MRP systems.

In 1990, Intergraph won an order to supply hardware and software for Sandia National Laboratories' Nirvana project, to implement a CAD/CAM system with a concurrent engineering capability.

Interleaf Inc.
Ten Canal Park
Cambridge, MA 02141

Product name(s) : Technical Publishing Software (TPS)

Founded in 1981, Interleaf shipped its first electronic publishing product in 1984. Today it is a multinational corporation with tens of thousands of users throughout the world. Interleaf integrates word processing, graphics, and page composition in one package. The basic version of Interleaf's WYSIWYG (What You See Is What You Get) TPS (Technical Publishing Software) can be augmented with optional modules. For the most sophisticated applications, Interleaf offers modules for MilSpec publishing, document management, advanced graphics, multi-page tables, and other specialized needs. TPS runs on computers such as Apollo, DEC, IBM, and Sun workstations, DOS-based 80386 machines and the Apple Macintosh II.

Interleaf offers a CALS Preparedness Package. The Preparedness Package is a combination of software, training, and support, designed to help users set up their own CALS environments and gain hands-on experience with the CALS interchange formats: SGML for text, IGES for engineering drawings, CCITT Group 4 for raster images, and 1840A for tape output.

The Preparedness Package gives users the opportunity to create CALS documents completely integrated with Interleaf's existing Technical Publishing Software product. Along with the software are customized CALS training programs and CALS technical support.

International Business Machines Corporation
400 Columbus Avenue
Valhalla, NY 10595-1396

Product name(s):

Data Communication Service (DCS)
Consolidated Data File (CDF)
CIM Communications and Data Facility (CIM CDF)
ProductManager: Engineering Management Edition
Product Engineering Support (PES)
Computer-Aided Design Integration (CADI)

IBM's CDF uses the IBM products DB2 and SQL/DS to consolidate, control and manage product definition data in relational format. IBM's DCS is a data communication facility that supports a distributed processing environment. It provides:

- Communication of data between users,
- Exchange and communication of data between different data bases,
- A single interface to the many product development applications, and
- A control and monitoring facility to track data communication requests and security violation.

The CIM Communications and Data Facility (CIM CDF) is IBM's product for managing data among CIM applications. It uses the IBM products DB2 and SQL/DS to store data in relational format.

The first of the CIM applications is the ProductManager: Engineering Management Edition. The ProductManager applications work to:

- Maintain the engineering product definition data (Product Structure Manager),
- Schedule and track engineering changes (Product Change Manager),
- Create and maintain routings (Routing Definition Manager),
- Estimate a product's cost, from design through product life cycle (Cost Evaluation Manager),
- Provide a link to the IBM System Application Architecture (SAA),
- Provide functions for electronic distribution of files,
- Interface to other systems (Product Data Interface),
- Assemble and route diverse collections of data (CIM Folder Manager), and
- Plan, schedule, and manage projects,

The Product Engineering Support (PES) and Computer-Aided Design Integration (CADI) were the initial programs in IBM's Engineering Management for CIM Family of Products. These programs provide bill of materials, engineering change, and electronic file functions with CADAM and CATIA products for users of IBM's mainframe System/370 VM environment.

The ProductManager: Engineering Management Edition and Computer-Aided Design Integration Version 2 programs are designed for the System/370 MVS environment.

International Computers Limited
Waterside Park
Cain Road
Bracknell, Berkshire RG12 1FA
London, England

Product name(s): Engineering Management Control System (EMCS)

The Engineering Management Control System from International Computers Limited (ICL) has six subsystems. These are written in a fourth generation language, making it easy to tailor and customize to user needs. The full data dictionary and all source code is made available. The system is currently implemented under ICL's VME operating system. ICL expects many EMCS users to require a high degree of data security. This is satisfied by VME's compliance to the U.S. DoD B1 classification for computer operating system security.

The six subsystems are:

- The Registry,
- The Bill of Materials Workbench,
- Engineering Change Control,
- Progress Monitoring,
- Project Tracking, and
- Interfacing to other Systems.

EMCS has been written with the expectation that it will be installed in multi-vendor environments. While the link to MRP is one-to-one and therefore relatively straightforward, the links to CAD/CAM/CAE are one-to-many and more complex. EMCS has facilities to control computer files on remote processors connected by Ethernet and TCP/IP, and to exercise controls over data in the files.

Image Systems Technology
165 Jordan Road
Rensselaer Technology Park
Troy, NY 12180

Product name(s):
CAD Overlay
CAD Overlay ESP
ViewBase

The CAD Overlay drawing conversion software allows users to build CAD data over a scanned image. CAD Overlay ESP (Edit, Save, and Plot) is a raster editor. It allows users to scan paper drawings, bring up the rasterized drawing on screen, erase what is no longer needed, update as required, and then plot the drawing. The scanned raster data can coexist with vector AutoCAD data on the screen. The raster and vector data can be edited, and merged into a single raster file. CAD Overlay ESP works with AutoCAD from Autodesk Inc. and VersaCAD Corporation's

VersaCAD software. PC-based ViewBase allows users to view indexed scanned drawings stored in a relational data base.

Ithaca Software
Ithaca, NY

Product name(s): Hierarchical object-oriented picture system (HOOPS)

Ithaca Software develops HOOPS, an object-oriented set of graphics subroutines.

Knowledge Based Systems, Inc.
2746 Longmire College Station
TX 77485

Product name(s):

AI0

AI2

Knowledge Based Systems products include AI0, an interactive PC-based tool for structured analysis, using hierarchical function modelling based on IDEF0 methodology rules, and AI2, an integrated PC-based system for informational analysis that incorporates the IDEF0, IDEF1, and IDEF1X methodologies.

KnowledgeWare Inc.
3340 Peachtree Road N.E.
Atlanta, GA 30326

Founded in 1979 by James Martin, KnowledgeWare Inc. develops the Information Engineering Workbench (IEW) CASE tool set. Front-end tools for planning, analysis, and design allow application requirements and specifications to be captured as diagrams. Back-end application generators can then process these pictures to create applications. All KnowledgeWare CASE tools share data through a common encyclopedia. In early 1990, there were more than 22,000 copies of KnowledgeWare CASE products in use.

Litton Industrial Automation Systems, Inc.
Integrated Automation Division
1301 Harbor Bay Parkway
Alameda, CA 94501

Integrated Automation/Litton provides integrated engineering document management solutions.

Matra Datavision
3 rue de la Terre de Feu
BP 246
91944 Les Ulis, France

Product name(s): Euclid-IS Data Management Facilities (EDMF)

Matra Datavision is a leading supplier of solids-based CAD/CAM and CAE systems. The company's major product, EUCLID-IS, is an integrated set of software programs that helps engineers to design, analyze, document, and manufacture complex mechanical and electromechanical products. EUCLID-IS is used on more than 1000 sites worldwide. The EUCLID-IS modeller provides the ability to define arbitrary solids and surfaces with the precision required for machining. EUCLID-IS and its application modules are built around the concept of a single integrated data base. Built-in data management tools are available with the EUCLID-IS Database module. These tools are used primarily within the development phase of the product design process.

Matra Datavision has developed EUCLID-IS Data Management Facilities (EDMF) to complement the built-in data management tools with those of a specialized data management system. This system would be used to manage the outputs of other CAD/CAM-related applications, and to manage the review/release and post-release phases of EUCLID-IS objects.

EDMF is the link between EUCLID-IS and the data management system. Its purpose is to make the two systems work together, and exchange and control each other's data with a well-defined protocol. The data management system used in the standard EDMF implementation will be DEC's EDCS II.

EDMF has two components: the EDMF kernel and the EDMF template. The kernel is a toolkit of extensions to the standard EUCLID-IS language, allowing complex manipulation of stored objects. It is supplied in binary form as a library of routines. The template is the source code of an application demonstrating use of the kernel. The template contains a EUCLID-IS interactive application to perform operations from a graphic workstation, and a forms-based application to perform some operations from an alphanumeric terminal with different screens for a manager and a user.

Microelectronics and Computer Technology Corporation
Austin, Texas

Microelectronics and Computer Technology Corporation (MCC) develops the Orion/Itasca object-oriented data base management system.

McDonnell Douglas
Manufacturing and Engineering Systems Company
P.O. Box 516
Saint Louis, MO 63166-0516
Product name(s): Product Structured Information Manager (Psi-Manager)

McDonnell Douglas, developer of the UNIGRAPHICS CAD/CAM system, offers Psi-Manager for data management. Psi-Manager provides a comprehensive environment for the generation, use, and management of engineering data. It provides a framework that draws together many existing islands of automation, interfacing discrete applications on an industry-standard OSI network.

Psi-Manager provides the means to capture and bring order to many pieces of information such as designer's notes, drafts, memos, specifications, requirements, and modification details. Online approvals and authorizations are provided to electronically manage the release process.

Product Structure Management is the core of the system and gives form to the model. Features of Psi-Manager include connectivity to modelling systems, versioning, product and process history, release management, and access control. With Psi-Manager, products, assemblies, parts, and the latest changes can be made available to authorized users.

Initial versions run on DEC platforms under VMS and on Hewlett Packard under UNIX. Although Psi-Manager is application-independent, its combination with modules from the UNIGRAPHICS library brings extra advantages.

Psi-Manager users include Eastman Kodak and McDonnell Douglas Helicopter Company.

Metier Management Systems
Metier House
23 Clayton Road
Hayes, Middlesex UB3 1AN
England

Product name(s):

Artemis
ECS 9000

Metier is a software engineering company specializing in project management systems. Through the Artemis project management systems available on mainframes, minicomputers, and PCs, it is a leading supplier of planning and control software. ECS 9000, previously named EMIS, is an engineering data management concept, oriented in the main towards linking Artemis to CADAM.

Northern Systems
The CAD/CAM Center, Riverside Park
Middlesborough, Cleveland TS2 1RJ
England

Product name(s): AURORA

Northern Systems was founded in 1987. It specializes in the areas of computer-aided design and manufacture.

AURORA is a document control and management system. It has three main modules:

- The Document Register is the base module for organizing drawings and documents and controlling user access.
- The Document Controller module maintains the relationships between documents, and provides the basic change control and sign-off facilities.
- The Design Manager module provides the required functionality for effective control of the design and related processes.

AURORA is a software system that can be implemented in single or multi-user mode, on PCs, large system clusters (including VAX), or mixed networks.

NovaCAD Inc.
MA

NovaCAD develops and markets products to address the drawing management problem. These include the NextCAD range of DEC Ultrix- and VMS-based image-capture and drawing management products.

Users include Fiat Automobile, Mutoh Kogyo, and Northrop Aircraft.

Octal Inc.
28280 Franklin Road
Southfield, MI 48034

Octal Inc. is a leading supplier of direct CAD/CAM data translators.

Ontologic Inc.
Billerica, MA

Ontologic develops the Ontos and Vbase object-oriented data base management systems. Ontos users include General Dynamics and McDonnell Douglas.

Optigraphics Corporation
9339 Carroll Park Drive
San Diego, CA 92121

Product name(s):
Optigraphics 3100 scanning and CAD conversion system
Optigraphics 4000 Document Automation System

OptiDRAFT station

Optigraphics Corp., founded in 1982, develops scanning, conversion, and document management systems. Functions addressed include:

- Document scanning (paper drawings/aperture cards),
- Raster edit,
- Raster-to-vector conversion (raster to CAD-compatible vector),
- Vector editing (of image converted vector data),

- Image and CAD drafting,
- Data output (plot/print),
- Data distribution (LAN & WAN),
- Image view/markup (PC-based),
- Document management (CURATOR, relational data base), and
- Document storage (optical disk / magnetic disk).

The OptiDRAFT station is a hybrid workstation that enables a user to make CAD-accurate revisions to an image, without converting the scanned document to vector format. It includes full-function 2-D CAD tools.

Optigraphics has an international customer base of approximately 200 users. These include Chrysler, ARCO, Martin Marietta, Lockheed Corp., TVA, Union Carbide, and the U.S. Navy.

Oracle Corporation
20 Davis Drive
Belmont, CA 94002

Oracle Corporation was founded in 1977 and introduced the ORACLE SQL relational data base management system in 1979. ORACLE is available on a wide range of computers, workstations, and PCs. It is used as a building block by many engineering data management system developers.

PAFEC Ltd.
Strelley Hall
Nottingham NG8 6PE
England

PAFEC Ltd. was founded in 1976, to develop and market a range of software tools for finite element analysis. PAFEC now offers a wide range of hardware independent software and complete turnkey systems. CAD modules include DOGS (Design Office Graphics System), SWANS, a sculptured surface modeller, and BOXER, a solid modeller.

RAVEN is an image scanning capture system. It accepts paper or film documents up to E-size (A0) at resolutions up to 400 dpi. The raster image can be edited, used as a background for CAD, or converted to vector format. LIONS is a drawing office management system. PAFEC Info-Manager is a fully-tailorable information management system for both graphical and non-graphical data.

Palette Systems Inc.
6 Trafalgar Square
Nashua, NH 03063

PALETTE is a suite of integrated graphic software modules, providing a central graphic data base; an interactive drafting program/graphic editor; a set of predefined interfaces to third-party systems, such as optical scanners, CAD, electronic publishing, word processing, and shop floor data collection; and a set of high-level graphic programming tools, allowing development and integration with other systems. Palette's Electronic Work Instruction (EWI) module automates engineering, manufacturing, and specification documents that contain text and graphics for instant, on-line access.

Prime Computers Inc.
100 Crosby Drive
Bedford, MA 01730

Product name(s):

EDM
EDMVault
EDMClient
EDMProjects
EDMProgramming

EDM software is a data management tool that provides complete data management, security, and project organization support for engineering applications. The major components of EDM are EDMVault, EDMClient, EDMProjects, and EDMProgramming.

EDMVault software provides complete data management services, including data storage, security/access control, back-up/recovery, report generation, and archive/restore capabilities. EDMVault performs the following functions:

- Check-in/check-out of files,
- Online catalogue and centralized storage, for many kinds of files,
- Access control for data in a distributed environment,
- Access to the data base restricted to authorized users,
- Modification to a given file is allowed for only one person at a time,
- Tools provided for archiving, restoring, backing up, and recovering files,
- Transfer of individual as well as sets of files and parts to and from the data base, and
- A method for relating a group of files and/or parts so that they can be treated as a single unit.

EDMClient software allows users or applications from CADDStation systems, VAX/VMS, or IBM/VM nodes to execute EDM commands from UNIX or within CADDS4X to file, retrieve, or manipulate data stored in the electronic document vault.

EDMProjects software provides project organization capabilities, including project definition and assignment, expanded revision/release control, and electronic sign-off. The various stages of work within each project can be defined. Customized authorities can be assigned to users. Users can be assigned to one or more projects with the same or different authorities for each project. User lists can be defined so that EDM alerts specified users when it is time to review a file, part, or file set. EDMProgramming software enables users to develop programs to service specific data management and project organization needs.

EDM functionality is provided for Sun SPARC, VAX/VMS, and IBM VM/SP systems. EDM uses a relational data base management system conforming to Structured Query Language (SQL) interface standards. EDM stores only control data in this data base. Control data allows EDM to maintain data integrity and perform product development control functions. EDM uses SQL/DS on IBM VM/SP systems, and ORACLE on VAX/VMS and Sun SPARC systems.

Initially marketed under the name PDM (Product Data Manager), EDM is used by companies such as General Dynamics, Otis, Ford, Rover, PSA, Philips, and Aeritalia.

Professional CAD/CAM Systems Inc.
9105 Guilford Road
Columbia, MD 21046

Professional CAD/CAM Systems Inc. has developed a PC-based system that accepts scanner output as input to provide full raster CAD editing and vector CAD editing, either separately or together. The system also provides a solution to storing and retrieving existing documents.

Quorom Computer Development Limited
18–19 Orton Enterprise Center
Bakewell Road, Orton Southgate
Peterborough PE2 0XU, England

Product name(s):
CADmaster
EDMS

CADmaster is an integrated package of hardware and software that links individual CAD workstations into a multi-user system, thus providing comprehensive management reporting, including drawing status and time reports. Quorum's EDMS is an integrated drawings, parts master, and multi-level bill of materials management system for engineers and drafters using PC-based AutoCAD systems.

Relational Technology Inc.
1080 Marina Village Parkway
Alameda, CA 94501

Product name(s):
INGRES
POSTGRES

RTI is a leading developer of relational data base management systems. The company's primary product is INGRES, a distributed SQL relational data base system. It is available on a wide range of computers and workstations. POSTGRES, an extension of INGRES, is an extended relational data base system.

RTI, renamed Ingres Corp., was purchased by ASK Computer Systems Inc. (Mountain View, CA) in 1990.

ScanGraphics Inc.
700 Abbott Drive
Broomall, PA 19008

Product name(s):

FRIENDS (Front-end Raster Imaging, Editing, Network, Display System)
RAVE Drawing Conversion System
CF Series Large Document Scanners

ScanGraphics was founded in 1972 to produce plotters. Later, it entered the scanner market and developed corresponding systems.

FRIENDS is an IBM PS/2-based data capture and document manipulation system. It includes the following modules:

- ScanServ—scanner control subsystem,
- ScanView—document viewing subsystem,
- ScanEdit—raster editing system, and
- ScanPlot—plot server.

RAVE is a raster-to-vector conversion system, running on IBM and DEC computers.

There are three models in the CF (Continuous Feed) Series of large document scanners for engineering drawings, technical illustrations, and other types of engineering documentation:

- CF 300, with three user-selectable resolutions at 100, 200, and 300 dpi,
- CF 500, with four user-selectable resolutions at 200, 300, 400, and 500 dpi, and
- CF 1000, with eight user-selectable resolutions at 200, 300, 400, 500, 600, 700, 800, and 1000 dpi.

Schlumberger Technologies CAD/CAM Division
4251 Plymouth Road
Ann Arbor, MI 48106

Product name(s): BRAVO 3

Schlumberger Technologies' CAD/CAM/CAE product BRAVO 3 uses DEC's EDCS-II for engineering data management. BRAVO 3 users can directly access EDCS-II to execute data management commands.

Servio Logic Corp.
Beaverton, Oregon

Product name(s) : GemStone

Servio Logic develops GemStone, an object-oriented data management system.

Sherpa Corporation
611 River Oaks Parkway
San Jose, CA 95134

Product name(s): Sherpa Design Management System (DMS)

Sherpa Corporation specializes in services and software that improve the design management process across diverse engineering and manufacturing environments, CAD/CAE applications, and hardware platforms. Sherpa's DMS software manages the design data and processes associated with a product from conception through revision, testing, approval, and release to manufacture. DMS provides for managing product information in a heterogeneous, distributed environment across wide area networks (WANs), as well as local area networks (LANs). Sherpa also markets application products that use Sherpa DMS. Application products address the diverse needs of an enterprise:

• Electronic vault for document management,
• Product structure manager for configuration management,
• Engineering request manager,
• Engineering change manager, and
• Master parts manager.

Sherpa DMS is available in heterogeneous DEC VAX, DEC Ultrix, Sun, HP, and Apollo environments.

DMS was introduced in 1986, and the first installation on a customer site was made that year. Among users of Sherpa systems are Ford Aerospace, Schlumberger Well Services, Hughes Radar System, Boeing Electronics, Baxter, Rockwell, and GPT Telecommunications Ltd.

Softsource
301 W. Holly
Bellingham, WA 98225

Product name(s):
DRAWING Librarian
BLOCK Librarian

The DRAWING Librarian is an AutoCAD Drawing-Slide-DXF display, control, and conversion program. It can display up to 25 AutoCAD drawings (DWG), slides (SLD), or DXF files on one screen. After viewing, one drawing can be automatically loaded to AutoCAD. Displayed AutoCAD drawings can be copied, erased, or converted to DXF.

The BLOCK Librarian is an AutoCAD graphic storage, retrieval, and control system. It allows users to select from up to 10,000 blocks or symbols.

SPICER Corporation
221 McIntyre Drive
Kitchener, Ontario
Canada N2R 1G1
Product name(s):
Viewpack
Turbo CAD Overlay

SPICER was founded in 1983. SPICER develops Image Management systems. At the heart of the system is a PC-based product called Viewpack. This provides viewing, markup, and editing of rasterized CAD drawings and scanned paper/aperture card images.

Boeing Computer Services, SPICER, and Tandem Computers have integrated the PPDM software with Viewpack. Turbo CAD Overlay operates in conjunction with AutoCAD and Viewpack, to provide complete CAD functionality in editing raster images.

SPICER clients include Allen-Bradley, Allied Bendix, Bell Helicopter, Bell Northern Research, Boeing, Cleveland Pneumatics, CR Industries, General Dynamics, General Electric, and General Motors.

Structural Dynamics Research Corporation
2000 Eastman Drive
Cincinnati, OH 45150
Product name(s): DMCS (Data Management and Control System)

SDRC provides engineering data management consultancy and services. They work with companies, assisting them in improving their design and management processes, and helping users to customize and implement DMCS.

DMCS manages and controls engineering data, both resident and nonresident, from concept to manufacture across heterogenous computer systems. It can operate at the departmental, project, and enterprise level, and can be interfaced directly with systems such as CAD/CAM, MCAE, CASE, MRP, and soon.

DMCS offers a flexible, customizable, rule-based system for managing the engineering process, and can provide the framework for CALS compliance. The functionality includes:

• Access control for data security,
• Automatic notification of actions in the design process,
• Approval process management,

- Change control,
- Data history maintenance,
- Status reporting,
- Configuration management,
- Life cycle management, and
- Archive management.

DMCS is available on DEC VAX, HP, Sun, Apollo, and IBM RS6000 and MVS systems.

Users of DMCS include Boeing Aerospace, Loral Space and Range Systems, Martin Marietta, Honeywell, Phillips, Rockwell, General Electric, Harris, Westinghouse, FMC, Iveco, Ford, Saab and Eaton.

Sybase Inc.
Berkeley,CA

Sybase Inc. develops the SYBASE relational data base management system.

Tandem Computers Inc.
19191 Vallco Parkway,
Cupertino, CA 95014

Product name(s):
Product and Process Document Management (PPDM)
Tandem Integrated Manufacturing Environment (T.I.M.E.)

PPDM is one of the three components of the T.I.M.E. architecture. PPDM includes core modules, application modules, and custom interfaces. The core modules include system administration, management of the PPDM central library, information management, and management of data flow between subsystems. The application modules control work authorization, change control, configuration management, archive management, release management, and management of external data. Customized system interfaces provide links between PPDM and CAD applications, as well as with other corporate and shop floor computers and devices.

Users of PPDM include Boeing Commercial Airplanes, Marconi Defense Systems, and Ontario Hydro. PPDM distributors in Europe include GEC Computer Services (UK) and Krupp Atlas Datensysteme (Germany).

TeamOne Systems, Inc.
2700 Augustine Drive
Santa Clara, CA 95054

TeamOne Systems, Inc. was founded in 1987, and has been delivering systems to support concurrent engineering environments since 1989.

The TeamNet concurrent engineering environment is targeted at product development teams in software, electronic engineering, and mechanical engineering. It contains a distributed object-oriented database, and provides for integrated functions: a distributed project repository, configuration management, change management and version control, and a database query system for reports. TeamNet is a UNIX-based product that can be hosted on any Sun workstation or server. It tracks design activity on any system in the network that runs NFS. Its user interface is based on X Window/Motif standards.

TeamNet is used at Martin Marietta Corp.'s Astronautics Division.

Teradata Corporation
El Segundo, CA
Teradata is a leading manufacturer of data base machines.

Texas Instruments
6550 Chase Oaks Blvd
Plano, TX 75023
Texas Instruments develops the Information Engineering Facility (IEF) CASE product.

Unic Systèmes
108, bd Richard Lenoir
75011 Paris, France
Product name(s): Calquer électronique
Unic Systèmes' product includes a high-resolution scanner, CAD Overlay software, CO-EDIT (to replace raster information by vectors), and CO-PLOT (to plot mixed raster/vector drawings). Unic Systèmes also markets ViewBase.

Unify Corp.
38970 Rosin Court
Sacremento, CA 95834-1633
UNIFY is a relational data base management system.

Unisys Corporation
P.O. Box 500
Blue Bell, PA 19424
Product name(s): Engineering Document Management System (EDMS)
The Unisys InfoImage Engineering Document Management System for digital storage of technical documentation enables users to integrate images of those documents with existing information processing systems. EDMS provides fully integrated, automated information management for documents, whether they are

on hard copy, in word processing files, or stored in a computer-aided design system.

The four basic functions of the system are:

- Electronic document storage, to manage documentation in multiple data bases on various media, and provide revision level control, as well as multi-level access security. The system retains a permanent record of documents on non-erasable, long-term optical storage media.
- High-speed optical scanners, to capture existing documents and enter them into the data base.
- Facilities for viewing, marking up, editing, and distributing engineering documents.
- Plot and print functions used to produce hard copy.

EDMS is targeted at enterprises that have a need for storage and recall of between 50,000 and 100,000 or more engineering documents of various sizes; a need for access to complete and accurate engineering information by multiple departments; long-term support requirements for products; multiple forms of engineering document storage, such as electronic, paper, and other media; and a need for improved document access control and security. EDMS uses the UNIX operating system-based U6000 computer, known as the document storage processor (DSP), to store and manage documents. EDMS supports a number of peripheral devices and workstations. The DSP is connected to an Ethernet TCP/IP local area network that can be integrated with the corporate network, to allow distribution of documents within a location and also to remote sites. EDMS works with the Unisys office automation system OFIS. Procedures can be built in OFIS, to distribute documents on a scheduled basis from one department to another.

The Document Database Management System (DDBMS) resides on the DSP and manages all interactions between the document data base and workstations on the network—accepting documents for storage, processing data base queries, and retrieving requested documents.

The DDBMS is a data base management application built on the ORACLE relational data base management system. Application functions include:

- User-friendly forms for data base queries and updates,
- Control of document storage and retrieval,
- Automatic document data base maintenance,
- Menu-based attribute definition and indexing,
- Full security through user access controls,
- Automatic back ups to optical disk,
- Audit trails and report generation for data base transactions, and
- Document distribution and management.

The EDMS data base consists of document images and a unique set of user-defined attributes, describing each document, such as document number, revision level, title, and release date. The attributes are used to retrieve specific documents or groups of documents from image storage. Multiple attributes can be defined, and related documents can be linked by means of these attributes, to create a particular set of information. Users can then query the data base and identify a group of documents relevant to a given task. Users can also maintain attribute information online for documents stored offline, such as hard copies stored in a file cabinet. It is also possible to establish indexes of the attributes most frequently used to search for documents. The DDBMS performs data base searches based on attributes entered on a query screen. When a user queries the data base, the DDBMS checks the index before searching the entire attribute data base. Access security is provided by the DDBMS with the capability to restrict access, storage, and modifications of both image and attribute data on either an individual or group basis. Direct printing from the data base is supported on a user-authorized basis.

The DDBMS provides an audit trail that allows the system administrator to track key transactions executed by the DDBMS and examine summary system statistics. A custom report generator facilitates status tracking with reports available on types of data base transactions, document usage, and frequency of data base transaction by user.

EDMS workstations are available on PC and UNIX platforms, and offer a range of functions:

- View—allows viewing and printing of documents.
- Markup—has all View capability, and in addition allows redlining to create a separate layer describing proposed changes, but does not alter the original document.
- Edit—has all of the above features, plus a raster editor to modify drawings and create new versions (or new drawings).
- Edit 2D—in addition to the above capabilities, has 2D CAD capability, vector-to-raster conversion, selective raster-to-vector conversion, and IGES and DXF import and export facilities.

EDMS systems are being used in Europe, Japan, and the United States.

Versacad Corporation
2124 Main Street
Huntington Beach, CA 92648

Versacad Corporation's VersaCAD is a leading PC-based CAD system. Versacad's Quick Render module uses HOOPS object-oriented display technology.

ViewStar Corp.
5820 Shellmound Street
Emeryville, CA 94608

ViewStar Corporation was founded and incorporated in 1986. Its ViewStar System is a distributed software system for management, distribution, and processing of multi-media electronic documents. It runs on 286, 386, and 486 PC workstations and operates on general purpose networks. The system has five major functional sub-systems:

- Document importation system (capture, conversion, and indexing of paper and electronically produced documents),
- Document presentation system (online display, annotation and modification of documents),
- Document management system (classification, storage, retrieval, and distribution of documents and related attribute data),
- Document workflow processing system (task definition, workflow modelling, and control),
- Application development platform (development environment for designing applications, user interfaces, data models, and workflow procedures).

ViewStar Corp. has customers in both the engineering and commercial sectors. Customers in the engineering/manufacturing sector include ARCO, Texaco, Public Service Electric & Gas, Proctor & Gamble, Georgia Power, Ralston Purina, Abbott Labs, and Westinghouse.

Wang Laboratories Inc.
One Industrial Avenue
Lowell, MA 01851

Wang is a leading worldwide supplier of integrated imaging solutions, with over 400 imaging systems sold. Wang's manufacturing customers include Allied Signal, Ford, Grumman, Hughes Aircraft, McDonnell Douglas, Raytheon, U.S. Steel and Westinghouse.

Wisdom Systems
29801 Euclid Avenue
Wickliffe, OH 44092

Wisdomwise Software is a family of products oriented towards simultaneous engineering. It includes the Concept Modeller. This presents the integrated knowledge, the rules, and the instructions needed to quickly and efficiently produce the product. Concept Modeller works with Prime Computervision and Parametric Technologies CAD packages. It runs on a wide range of hardware, including Sun, HP, Apollo, Symbolics, and IBM Unix workstations.

Xerox Corporation
P.O. Box 24
Rochester, NY 14692

Xerox's Ventura Publisher is a leading page composition/technical publishing system. It accepts data in several standard CAD exchange formats and from several word processing systems.

3M, 3M Center
St Paul, MI 55144-100

3M's Doculink 7500 provides functions for scanning, modifying, plotting, and distributing engineering documents. Based on industry standards (UNIX, Ethernet, Oracle), it can be linked to a variety of scanners and CAD systems.

Appendix 3

Product Names

PRODUCT NAME, TRADEMARK, OR REGISTERED TRADEMARK	COMPANY NAME
ADICAD	ADI Software GmbH
Adra Vault™	Adra Systems, Inc.
AIO™	Knowledge Based Systems Inc.
AIX™	International Business Machines Corporation
Apollo™	Apollo Computer Division of Hewlett Packard Corporation
ARTEMIS™	Metier Management Systems
AURORA	Northern Systems
AutoBASE™	Cyco International
AutoCAD™	Autodesk Inc.
AutoEDMS	ACS Telecom
AutoManager	Cyco International
AutoSIGHT™	AutoSIGHT Inc.
BLOCK Librarian™	Softsource
BOXER	Pafec Ltd.
CAD Overlay™	Image Systems Technology Inc.
CAD Overlay ESP™	Image Systems Technology Inc.
CADAM™	CADAM Inc.
CADEX™	Database Applications Inc.
CADDStation™	Computervision Corporation

CADI	International Business Machines Corporation
CADMAC	Infodetics Corporation
CADMAN	Cadco Ltd. / SMC Ltd.
CADRA-III™	Adra Systems, Inc.
CATIA™	Dassault Systèmes
CIM CDF	International Business Machines Corporation
Composite Image™	Auto-trol Technology Corporation
Concept Modeller™	Wisdom Systems
CURATOR™	Optigraphics Corporation
DB2™	International Business Machines Corporation
DCS	International Business Machines Corporation
DDBMS	Unisys Corporation
DDTMS	Alpharel
DEC™	Digital Equipment Corporation
DECnet	Digital Equipment Corporation
DECwindows	Digital Equipment Corporation
DFS	Access Corporation
DMCS	Structural Dynamics Research Corporation
DME™	Context Corporation
DMS™	Sherpa Corporation
DOGS	Pafec Ltd.
DORIS	ASL
DRAWBASE™	CADworks Inc.
DRAWING Librarian™	Softsource
DXF™	Autodesk Inc.
ECC™	ACCESS Corporation
EDCS-II	Digital Equipment Corporation
EDICS™	Access Corp.
EDL	Control Data Corp.
EDM	Computervision Corporation
EDMClient	Computervision Corporation
EDMF	Matra Datavision
EDMProgramming™	Computervision Corporation
EDMProjects™	Computervision Corporation
EDMVault™	Computervision Corporation

EDMS	Quorum Computer Developments
EDMS	Unisys Corp.
EDS	Electronic Data Systems Corp.
EIMS™	Auto-trol Technology Corporation
EMCS	International Computers Limited
EMPRESS™	Rhodnius, Inc.
EM*MACS	Infodetics Corporation
EP-SDV	Eigner & Partner GmbH
ERIC	ASL
Ethernet	Xerox Corporation
EUCLID-IS	Matra Datavision
EWI (TM)	Palette Systems
EXCELERATOR	Index Technology Corporation
FastTrax	E. I. DuPont de Nemours & Co.
FORMTEK	Lockheed Corp.
FRIENDS	Scangraphics Inc.
Gédéon	Archivage Systèmes
HOOPS	Ithaca Software
HPGL	Hewlett Packard Corporation
IBM	International Business Machines Corporation
ICEM	Control Data Corporation
IEF	Texas Instruments
IEW	KnowledgeWare Inc.
ImageMaster™	Cimage Corporation
Impression™	Eyring Inc.
InfoImage™	Unisys Corporation
INGRES™	Relational Technology, Inc.
Intelligent Documentation	CIMLINC
Interleaf™	Interleaf Inc.
ISIF	Intergraph Corporation
I/PDM	Intergraph Corporation
LIONS	Pafec Ltd.
Macintosh™	Apple Computer, Inc.
Motif™	Open Systems Foundation
MVS	International Business Machines Corporation

Mylar™	E.I. DuPont de Nemours & Co.
NetWare™	Novell Inc.
NextCAD	NovaCAD Inc.
NFS™	Sun Microsystems Inc.
Novell	Novell Inc.
OFIS	Unisys Corporation
OptiDRAFT™	Optigraphics Corporation
Oracle™	Oracle Corporation
PageMaker	Aldus Corporation
Personal EDMS	ACS Telecom
PES	International Business Machines Corporation
PostScript™	Adobe Systems Inc.
PPDM	Tandem Computers Inc.
PrePARE™	Rosetta Technologies Inc.
PreVIEW™	Rosetta Technologies Inc.
Quick Render™	Versacad Corporation
RAVE™	Scangraphics Inc.
RAVEN	Pafec Ltd.
RETREEVE	Auto-scan Systems Inc.
SAA™	International Business Machines Corporation
ScanEdit™	Scangraphics Inc.
ScanPlot™	Scangraphics Inc.
ScanServ™	Scangraphics Inc.
ScanView™	Scangraphics Inc.
Sherpa™	Sherpa Corporation
SNA™	International Business Machines Corporation
SQL/DS	International Business Machines Corporation
SUN™	Sun Microsystems Inc.
SWANS	Pafec Ltd.
Sybase™	Sybase, Inc.
TDMF™	Electronic Data Systems Corporation
TDMS	Access Corporation
TeamOne™	TeamOne Systems Inc.
TeamNet™	TeamOne Systems Inc.
Teamwork	Cadre Technologies Inc.

Tech Illustrator™	Auto-trol Technology Corporation
Tech Image™	Auto-trol Technology Corporation
The Integrator™	C-TAD Systems Inc.
TIM™	Auto-trol Technology Corporation
TIME™	Tandem Computers Inc.
TIPS	CAYLX Software Ltd. / TRIO Software
TPS™	Interleaf Inc.
UNIGRAPHICS	McDonnell Douglas
UNIX	American Telephone & Telegraph
VAX™	Digital Equipment Corporation
Ventura Publisher	Xerox Corporation
VersaCAD™	Versacad Corporation
ViewBase™	Image Systems Technology Inc.
ViewStar™	ViewStar Corporation
VM	International Business Machines Corporation
VME	International Computers Limited
VMS™	Digital Equipment Corporation
Wisdomwise Software™	Wisdom Systems
WorkFlo™	FileNet Corporation
X Window System™	Massachussetts Institute of Technology

All other brand and product names are trademarks or registered trademarks of their respective companies.

Appendix 4

Acronyms

AFNOR	Association Française de Normalisation
ANSI	American National Standards Institute
ASCII	American Standard Code for Information Interchange
BS	British Standard
CAD	Computer-Aided Design
CAD/CAM	Computer-Aided Design/Computer-Aided Manufacturing
CAE	Computer-Aided Engineering
CALS	Computer-Aided Acquisition and Logistic Support
CAM	Computer-Aided Manufacturing
CAP	Computer-Aided Publishing
CAPP	Computer-Aided Process Planning
CASE	Computer-Aided Software Engineering
CCITT	Consultative Committee for International Telegraphy and Telephony
CGM	Computer Graphics Metafile
CIM	Computer Integrated Manufacturing
CIM-OSA	Computer Integrated Manufacturing Open Systems Architecture
CSG	Constructive Solid Geometry
DBMS	Data Base Management System
DD	Data Dictionary
DDL	Data Definition Language
DFA	Design for Assembly
DFM	Design for Manufacture

DIN	Deutsches Institut für Normung
DML	Data Manipulation Language
DoD	Department of Defense
EC	Engineering Change
ECN	Engineering Change Number
ECO	Engineering Change Order
ECR	Engineering Change Request
ED	Engineering Data
EDI	Electronic Data Interchange
EDIF	Electronic Design Interchange Format
EDM	Engineering Data Management
EDMS	Engineering Data Management System
EDM/EWM	Engineering Data Management/Engineering Workflow Management
EDP	Electronic Data Processing
EIM	Engineering Information Management
EP	Electronic Publishing
ESPRIT	European Strategic Program for Research and Development in Information Technologies
EW	Engineering Workflow
EWM	Engineering Workflow Management
F&A	Finance and Administration
FDDI	Fiber-Distributed Data Interface
FMS	Flexible Manufacturing System
FTAM	File Transfer Access and Management
FTP	File Transfer Protocol
GB	Gigabyte
GKS	Graphic Kernel System
GMP	Geometric Modelling Project
GT	Group Technology
GUI	Graphic User Interface
HDSL	High-Level Data Specification Language
ICAM	Integrated Computer-Aided Manufacturing
IEEE	Institute of Electrical and Electronics Engineers
IGES	Initial Graphics Exchange Specification

IPAD	Integrated Program for Aerospace Vehicle Design
IPO	IGES/PDES Organization
IS	Information Systems
ISDN	Integrated Services Digital Network
ISO	International Standards Organization
IT	Information Technology
JIT	Just-In-Time
KB	Kilobyte
LAN	Local Area Network
MAP	Manufacturing Automation Protocol
MB	Megabyte
Mbps	Megabits per second
MCAE	Mechanical Computer-Aided Engineering
MIS	Management Information Systems
MIT	Massachussetts Institute of Technology
MOD	Ministry of Defence (UK)
MRP 2	Manufacturing Resource Planning
NIAM	Nijssen Information Analysis Method
NBS	National Bureau of Standards
NC	Numerical Control
NIST	National Institute for Standards and Technology
ODA	Office Document Architecture
ODETTE	Organization for Data Exchange by Teletransmission in Europe
ODIF	Office Document Interchange Format
OSI	Open Systems Interconnection
PDDI	Product Definition Data Interface
PDES	Product Data Exchange using STEP
PHIGS	Programmer's Hierarchical Interactive Graphics System
PSDN	Packet Switching Data Network
QA	Quality Assurance
QC	Quality Control
R&D	Research and Development
SC	Subcommittee
SDIF	SGML Document Interchange Format
SET	Standard d'Echange et de Transfert

SGML	Standard Generalized Mark-up Language
SQL	Structured Query Language
STEP	Standard for the Exchange of Product Model Data
TB	Terabyte
TC	Technical Committee
TCP/IP	Transmission Control Protocol/Internet Protocol
TIFF	Tagged Image Format Files
TOP	Technical and Office Protocol
TP	Technical Publishing
TQM	Total Quality Management
TRIF	Tiled Raster Interchange Format
VDA-FS	Verband der Automobilindustrie—Flächen Schnittstelle
VDA-IS	Verband der Automobilindustrie—IGES Subset
VHDL	VHSIC Hardware Description Language
VHSIC	Very High Speed Integrated Circuit
WAN	Wide Area Network
WIMP	Window, Icon, Menu, Pointer
WIP	Work In Process
WYSIWYG	What You See Is What You Get
XBF	Experimental Boundary File

Index